International Directory of
Model Farm Tractors

by Raymond E. Crilley Sr. and Charles E. Burkholder

Schiffer Publishing Ltd

Box E, Exton, Pennsylvania 19341

Acknowledgments

The authors wish to express their sincere appreciation to anyone involved with the research and production of this book particularly to those listed below for all the extra effort afforded this project:

Bill Bartlett Jr., collector from Illinois; Wayne Beckmom, collector from Indiana; David Bate, collector from Albrighton, England; Jon Bright, collector from Indiana; Milford Cooper, collector from Illinois; Wayne Cooper, collector from Pennsylvania; Robert and Judy Condray, collectors from Kansas; J. Peter Crilley, computer operator from Pennsylvania; Laura English, secretary from Pennsylvania; Jean-David Geyl, collector from Sanary, France; Richard Havens, photographer from Michigan; Verlan Heberer, collector from Illinois; Ronnie Johnson, collector from Minnesota; David Kerby, collector from Ohio; Graham and Michelle Miller, collectors from Herts, England; Keith Oltrogge, collector from Iowa; Wayne Samuelson, collector from Iowa; Gilbert Sontag, collector from Vulmerange-Les Mines, France; Eldon Trumm, collector from Iowa; A.H. van Hoorn, collector from Norgerweg, Netherlands; Bob Weaver, collector from Pennsylvania; Franco Zampicinini, collector from Torino, Italy; Bob Zarse, collector from Indiana; and Fabio Zubini, collector from Trieste, Italy.

Some of the models photographed and labeled in the book have had decals added to the original model.

Front cover: 031503-006, 040521-025, 100400-162, 090803-100A, and 112102-006
Back cover: PT1512-001, PT0301-007, and 061719-054
Spine picture: 061718-015A

Contents

The New Farmall Cub Tractor Assembly Kit

Everybody, young or old, likes to build things—to put things together.

Here is a Farmall Cub tractor that comes knocked down, ready for any clever youngster to assemble. Grownups will be fascinated by the job also. The task is not difficult and when completed the Little Farmall Cub is as lifelike as the full-size original tractor.

Most of the unit is made of bright Farmall red plastic. The steering gear and certain other more delicate parts are made of steel to provide durability.

When assembled the tractor measures 6 inches long; 4 inches high; and 3 inches wide. Cutaway carton as shown here measures 6 x 10¼ x 2. One dozen cartons are packed in a strong corrugated container for shipment.

A-499-KK

THE NEW FARMALL CUB TRACTOR ASSEMBLY KIT A-499-KK

JUST WHAT YOU ASKED FOR CR-1070-E 10-3

PLASTIC TD-24 TRACTOR MODELS

Here is the famous remote control electrically operated TD-24 with blade unit. Operator plugs into AC 110 outlet and can manipulate model forward, reverse, make it pivot or turn. It has two complete speed ratios. Supplied complete ready to use with 14 foot of cord and control box.

Crawler also supplied less blade. All units of durable red molded plastic, 13½" long and 8" wide.

TD-24 unit with bullgrader blade has long been popular in the IH toy field.

Other IH toy models not illustrated in this folder available include Farmall "450" tractor, plows, discs, loaders, wagons spreaders and Child's riding tractorcycle may be ordered on AD-144RA.

INTERNATIONAL HARVESTER COMPANY
180 NORTH MICHIGAN AVE. • CHICAGO 1, ILLINOIS

CR-338-6   5-3

PLASTIC TD-24 TRACTOR MODELS CR-338-6 5-3 col.

Now Available!
Miniature Plastic Farmall Tractors

Profit Maker . . .
Good Will Builder . . .
Display and Sell
Every Day . . .

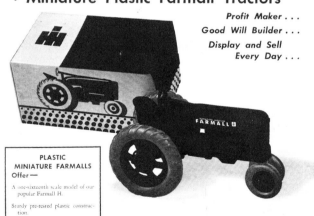

PLASTIC MINIATURE FARMALLS Offer—

A one-sixteenth scale model of our popular Farmall H.

Sturdy pre-tested plastic construction.

Harvester red color in plastic, no paint to chip.

Rubber tires mounted on wheels, molded as separate units.

Special drawbar to provide for pulling other units.

Attractively boxed for display and sales purposes.

A new advancement in product miniatures at a low price.

EVERYONE A PROSPECT!

During a period of several months test sale by Harvester dealers this item has proved its popularity. Not only can it be sold in quantity for toy use but many thousands will be bought by hobbyists and other interested persons. Based on present prices, it can be sold profitably at $1.50 retail.

Dealers will find that it pays to keep these tractors on display in their showrooms and parts departments. They will bring in additional sales by attracting customers into the store.

An Alert Dealer Will Keep These In Stock and On Display

ORDER FROM YOUR BRANCH TODAY - - - ALL ORDERS IN LOTS OF
DOZENS OR MORE - - SHIPMENTS BEING MADE NOW BY OUR SUPPLIER

MINIATURE PLASTIC FARMALL TRACTORS A-121-KK 2-4

Introduction to Model Farm Tractor Collecting 1

An oversimplification of the question, "Who collects farm toys?", results in an answer "Almost anyone". However, it is necessary to go into detail as to "Why do people collect farm toys?" to fully understand who these people really are.

The rather obvious conclusion that comes to mind would be that farmers collect farm toys, and, this conclusion has much validity. In fact, in the very early days of farm toys during the 1920's and 30's, farm children were sometimes treated with a model farm tractor, just like dad's, by the farm machinery dealer when dad purchased a real tractor. This toy would soon become a keepsake, for luxuries such as toys were scarce during the depression years. A few years after receiving the toy, a farm child soon outgrew playing and had more than sufficient work to occupy his time. The toy was relegated to the attic or some other area for safe keeping.

In mentioning farm machinery dealers, it is easy to understand a vested interest in farm toys on their part. It was soon realized that these farm toys were very positive advertising pieces. A youngster soon learned to identify with a particular brand of tractor, be it green, red, gray or another color, because of the toy model. Later in life, this youngster was introduced to that brand when purchasing a real tractor of his own.

Before the days of the modern communications such as radio, television, newspapers, telephones, etc., companies would employ traveling salesmen. In many cases, these salesmen would use a miniature, either a salesman's sample or a toy model, to illustrate the virtues of the product being marketed. Naturally, farm children would be intrigued by these miniatures as a sale was being consumated over the kitchen table. Imagine the delight of that child when a toy model was left behind.

While on the subject of salesman's samples, it must be mentioned that these miniatures were far more intricate than toys and were too expensive to leave behind after the sale. The toy models were much less expensive than the salesman's sample, and therefore, better served the need.

Anyone who spent untold hours doing farm chores using a particular tractor soon developed a like or dislike for it. One in that position can soon relate the unique peculiarities a particular tractor possesses. For example, the old Fordson F models were notoriously difficult to start in cold weather. Even with the addition of "hotshot" batteries, as well as a wide variety of other modifications, starting remained a major effort on the part of the operator. The number of dislocated thumbs and broken wrists from the crank kicking back while attempting to start the engine can only be estimated, but surely it is a staggering figure. Having suffered from the crank kickback, one would always remember that tractor, and not in a kindly fashion. On the other hand, there are many fonder memories associated with farm work, particularly during the harvest season. Harvesting was a major event, particularly during the grain threshing season. Even though it was hard work, it was a time when neighbors would work together and at mid-day or evening sit down to a real farm style meal, from which no one left hungry. After the meal, time was taken to relax and talk about everything and anything. Of course the "old timers" would relate stories about times before modern equipment such as tractors. It was during these times that stories were told about horses being the mainstay in performing farm chores. And, not unlike tractors, horses had unique personalities or peculiarities. But all the horse stories withstanding, the farm youngsters knew that horses were doomed to easier days of retirement because everyone knew that tractors did not require hay and grain when not being used. And, tractors did not produce manure which required removal, adding to the chores needing done on the farm.

Arcade cast iron teams and steel wheel spreader, Weber wagon and rubber tire spreader, all representing McCormick-Deering models.

Early toy makers such as Arcade and Vindex included teams of horses among their offerings. Arcade's McCormick-Deering spreader and Weber wagon featured a very nice cast iron team of horses simulating the pulling force. During the depression, the Vindex Company, whose main business was manufacturing sewing machines, changed to toy making as a means of keeping the employees busy when the demand for sewing machines dropped off. During this period, they produced a wide variety of toys for the John Deere Company including some horse-drawn implements which were pulled by cast iron horse teams. Arcade and Vindex teams were similar in size but could be easily identified because the Arcades had a simulated wheel cast on each horses' one front foot while the Vindex teams had a ball cast on the feet.

Wilkens cast iron horse and sulky plow.

Before either Arcade or Vindex became the leaders in farm toy industry, another manufacturer, Wilkens, made some very fine miniature implements, horse-drawn of course, including a mowing machine, a hay tedder, a dump rake and a sulky plow. These creations by Wilkens were of slightly larger scale than the Arcade and Vindex models. All of these are very desirable collector pieces today.

Wilkens cast iron horse and dump rake.

Older people who remember the horse as a power source on the farm are more likely to be collectors of the implements and teams, although this is not to imply that this is a rule.

This brings up the topic of "period" collecting. It appears as though many people who become farm toy collectors are interested in a particular segment of time relating to the content of their respective collections. For example, the older collectors like the horse-drawn implements or perhaps the steam era miniatures which includes steam traction engines, water wagons and threshers. While a few miniatures were made back then, the steam traction engines really became popular with collectors during the 50's and 60's. This is when the Brubakers, Petersons, Whites, etc. began producing the cast aluminum outfits which are still being produced today by Don Irvin. Robert Gray made some very nice korloy models of steam traction engines and threshers during the 60's and 70's. Incidentally, korloy is a material somewhat like cast iron, but non-metallic. It is re-cycled from old discarded parking meters.

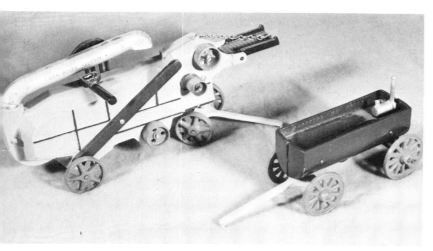

Robert Gray Red River Special thresher and water wagon.

Younger collectors, or shall we say, middle aged collectors, may be more interested in miniatures representing real tractors produced in the 1940's or 50's. One might theorize that a collector's interest traces back to his formative years when he was growing up. Frequently this is the period of time that made the greatest impression on him. So now, being older, and wiser, he looks back and very possibly, appreciates a little better, those early years.

The youngsters, and this may be defined as anyone younger than ourselves, could well identify with tractors of the 60's and 70's. For this is the time that farm toys really came into great prominence. The Ertl Company captured the industry and expanded from their production of John Deeres and Internationals during the middle and later 1950's to other brands including Allis-Chalmers, Case, Minneapolis-Moline, Massey-Ferguson and Oliver. It was then that the "standard" size became the 1/16 scale. Of course, Hubley continued the large 1/12 scale Fords which was perported to dispell the "small tractor" image Ford was attempting to rid itself of.

Rick McAllister's 1/32 model farm.

Now comes the younger generation who thought the 1/16 and 1/12 scale tractors were nice, but really took up a lot of room. So why not collect 1/32 or 1/64 scale tractors which required much less room to display or store? Now the 1/64 scale lines are the rage. The big three of the 80's, the Ertl Company, Scale Models, and Mini Toys all produce a variety of 1/64 scale models here in the United States. The first two companies mentioned here also produce 1/32 scale models as do many foreign producers such as Britains Ltd., Hausser-Elastolin, Siku, Yaxon and others.

Rick McAllister's 1/32 model farm.

Both the 1/32 and 1/64 scales, as well as 1/43 and 1/50 scales all lend themselves to creating farm scenes or dioramas. Many collectors of these smaller scales like to set up a farm scene adding other enhancements such as buildings, silos, grain storages, animals, fences and many others. To this end, some even produce an intricate miniature of the "home farm", adding a particularly personal touch to the hobby.

David Preusse's model I.H. dealership.

Some collectors prefer to stay with a particular brand such as John Deere or International Harvester. There is usually a particular reason to collect just one line, and that reason may be that the line represented happens to be the same as the real machinery he uses on the farm, or sells, or otherwise has an involvement. It is not uncommon to see a collection of several hundred pieces, all the same color such as green and yellow. This, of course, represents the John Deere line, which, by far, is the most popular line in the United States. A collector in Canada may have all Massey-Harris and Massey-Ferguson since these lines were and are quite popular there. A collector in England may favor the David Brown line while a collector in Italy may have all Fiats.

The interesting part of the hobby is that a collector can individualize his collection by adding what appeals to him either in scale, brand or period of time. Except for some very new collections, each collection will be different and somewhat unique.

At this point, it is necessary to digress a bit so that a reader does not get the idea that farm toy model collecting is limited to men and boys, although we have all seen the sign "The difference between men and boys is the price of their toys." Presently, there are a fair number of ladies involved in the farm toy model hobby, not just as a partner as in a husband-wife relationship, but as an individual having her own interests. For the non-farm readers, it must be mentioned that farm ladies are very much involved in farm chores including the field work particularly at planting or harvesting times. Many farm ladies welcome the opportunity to get out of the home and operate the tractor or combine. I personally have had grain combined by a lady operating the harvester. Just as in caring for the livestock, ladies take the extra minutes to be thorough and do the job well. Many of the reasons, or if you will, excuses, for collecting farm toys, are the same for both the ladies and the gentlemen.

Collecting can be somewhat contagious. Once a person has had the opportunity to see a few collections, and realize the fun involved, the "bug" bites, and it grows from there. A novice collector then discovers that there are publications such as Miniature *Tractor & Implement*, which specialize in farm toy collecting. Through the publication, he can learn more about old and new models, variations of particular models, where and when the shows and auctions will be held, where to buy both new and obsolete models, and probably most important, the names and addresses of other collectors who share his desire to acquire a respectable collection.

It is at this point a novice can be overwhelmed at the scope of the hobby. He sees so many things in which to spend his money and realizes that there is only so much money to spend. Being a prudent person he must make a decision to put constraints on his desires and be realistic regarding his expectations. There are so many nice models, available that the heirs to Howard Hughes fortune might be able to purchase all of them, but all of us are not that fortunate to have a limitless money supply for wants and desires, not to mention necessities. So after this initial awareness of reality, the novice has some readjusting to do. He may decide that he wants to narrow down his field of endeavor and specialize. This is good in that he now has some "trading stock" or goods to sell, enabling him to barter some models, exchanging some he doesn't want for others he does want.

The best way to accomplish this is to attend a farm toy show. One cannot think of farm toy shows without mentioning the grandaddy of farm toy shows, the National Farm Toy Show founded by the Toy Farmer, Claire Scheibe. Although there were earlier small regional farm toy swap meets and shows, Claire's promotion resulted in the first truely national show. His first show was held in 1978 in Dyersville, Iowa home of the Ertl Company, and Scale Models, both important farm toy manufacturers.

This first National Farm Toy Show was held in St. Francis Xavier's School in downtown Dyersville. The exhibitors had an opportunity to deal with each other, then it was opened up to the public which was composed of many collectors and many curious "lookers". Interestingly enough, many of the "lookers" came back the next year with an armful of farm toys to sell, and some even emerged as exhibitors at succeeding shows.

As part of the promotion for the first show, Claire offered a limited production "show tractor". This show tractor was a re-release of the Ertl Farmall 560, but with some modifications including the addition of a wide front axle. This initial offering was so popular that the show tractor became an annual part of the show promotion. The Toy Farm Show tractors are featured in another part of this book.

Another feature of the first National Farm Toy Show was a farm toy auction. Here exhibitors could consign models which were to be auctioned to the highest bidder. Veteran farm auctioneer Wally Hooker of Indiana entertained the audience with a few stories, then got down to business, selling farm tractor models, much the same way he sold the real ones. The auction was an instant success and became a permanent feature of the show.

The show has grown steadily, outgrowing its original site, so that it had to be relocated to Beckman High School on the South side of Dyersville. During a recent show with over 10,000 visitors, this new location was hardly adequate.

The number of farm toy shows had proliferated and now totals well over one hundred. This growth has spread in all directions from Dyersville, Iowa, and new shows are continually being organized.

Attending one of the major shows is an excellent way of becoming familiar with the prices that farm toy models are bringing. One can brouse among the dealers' tables and observe the prices on various items of interest. It is not at all unusual to see a variety of prices for an item on different tables. So, if you are looking for a particular item, it would be wise to "shop around."

While on the subject of prices, it should be noted that farm toy auctions frequently have a profound influence on farm toy prices. For example, if a Vindex John Deere "D" brings $900 at an auction, don't be surprised to find dealers' price tags reflecting a similar price. A caution is in order at this point. Auctions can be very emotional situations and bidders can go higher than they intended to bid, and possibly regret it later. It has also been known that a consigner will get involved in the bidding on his own item and deliberately run up the bid. Although this is an infrequent situation, it can artificially inflate the price of that item.

David Nolt of Pennsylvania recognized the problems mentioned above, and further realized that collectors and dealers wanted some type of a price guide for farm toys. So he set up a program on his home computer which would input various auction prices, eliminate any particular highs or lows for a specific item, then average the prices from many auctions. His Farm Toy Price Guide is periodically updated so that up and down trends can be programmed, thus giving a current price average for each item.

As one might suspect, prices are determined by many factors, the most important one being the general farm economy. When farm prices are up, so too are prices of farm toys and when prices are down, farm toy prices tend to fall.

Another very important pricing factor is condition of the toy. Frequently, a "new in the box" toy will bring two, three, four or more times the price of a "played with" model. There seems to be a great attraction for the new-boxed items, particularly if that item is really old, for example, from the 1950's. Another chapter in this book deals with farm toy tractor model boxes and packaging.

With all the discussion about prices, it should not be surprising that this hobby can become an expensive one, depending upon the tastes of each collector. Some collectors make a decision to get involved in the hobby to the extent that the hobby helps support itself.

One way this is done is by becoming a dealer. Purchasing extra models, then reselling them at a profit is one type of involvement. Another way of making a profit is to provide a service for other collectors such as repairing, restoring or repainting used models. A few collectors have established full-time occupations doing such work. If the quality of work is very good, business will grow. There is always need for replacement or dress-up decals. A few collectors have gone into business manufacturing decals, and do a thriving business. Other collectors have gone into business customizing models, taking stock models and converting them into rather interesting custom miniatures.

Still others have gone even further setting up manufacturing shops where they make models which were never commercially manufactured previously. Several of these manufacturers are listed elsewhere in the book and photographs of their products appear in the "models" chapter. Obviously, quality of these small time manufacturers' products will vary considerably, some being very well detailed and accurate while others are rather crude. Generally, prices charged for the models will reflect the quality somewhat, although this is not always necessarily the case.

We have reviewed some of the reasons that make people want to become involved with farm toy collecting. To be sure, there are many other reasons why a collector might "join the crowd" and start his own unique collection of farm toys.

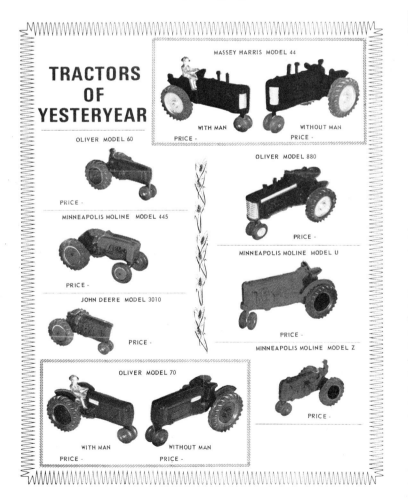

TRACTORS OF YESTERYEAR—Jim Hosch.

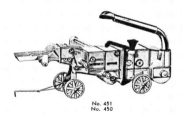

Farmall tractors in THEY LOOK REAL—Arcade ad.

8

Displaying the Model Farm Tractor Collection 2

One of the enjoyments of collecting something, is to be able to easily view your acquisitions. Visitors and others will also receive enjoyment from viewing nicely displayed models.

Having an organized arrangement of the collection aids in locating selected models easily and quickly. Rummaging through a pile of boxes can consume unnecessary time and cause frustration when desiring to locate a specific model for whatever reason; checking out a variation detail; determining if a certain model is in your collection, etc.

Display Arrangements

Several arrangement categories may be used to display ones' models. The selection of the catagory will be determined by the collector. He may select one or more of the following methods:

1. Brands—Case; Ford; John Deere; etc.
2. Construction material—plastic, cast iron, die-cast, etc.
3. Manufacturers of the model—Ertl, Scale Models, etc.
4. Scale or size—1/64, 1/32, 1/16, etc.
5. Type of machine—all manure spreaders displayed in a group, plows, wagons, etc.
6. Implements displayed hitched to or mounted to corresponding tractor. For example: two row John Deere mounted corn picker attached to a "60" type tractor, pulling a flare box wagon.
7. Construction models—vs—farm models.
8. Foreign—vs—domestic made models.

Nearly every collector has a different "twist" to displaying their finds. Visiting another collectors' display can be very interesting to see familiar models displayed in a different way.

Remember, the primary person to satisfy when displaying a collection is the collector himself.

Choosing a Method of Displaying Your Models

There are several factors to consider in selecting how to display your collection. They include:

1. Cost
2. Available space
3. Visability
4. Security

Costs

Providing an attractive display of your collection can be costly. Many collectors have limited finances and may prefer to invest their extra funds in adding to their collection rather than building fancy display units.

One must first determine his goals and how much income they wish to devote to collecting and displaying their hobby. Collections can be displayed very attractively with only a small amount of costs if a little imagination is used as you can see in the selected pictures. On the other hand, large sums of money can be spent on custom made cabinets, display units having special lighting, dust-free glass enclosures and costly hardwood cabinetry.

Available Space

Some homes are very limited in having extra space to display moderately large collections. Often a spare room, or basement are converted into a display area, especially after the collection has grown. It may have started out with a few models on top of the fire place mantel or on a nick-nack shelf or china cabinet—if the wife allowed it—but usually it's not long until more space is needed. More fortunate collectors have built on an extra room to the home just for displaying the collection. Others have converted an outbuilding into an area to store and display their collection. The office of a professional or business person may be seen with shelves displaying a collection.

Visability

To be most appreciated by the collector and visitors, the display should be as visible as possible. Obstructions, such as a door or window dividers used to hold glass in front of shelving, tend to distract from the clear view of models on display. Open shelves or shelving with unobstructed glass fronts, offer a clean, clear view. Proper lighting, preferably indirect lightening, helps in bringing out the detail of various models. Glass shelves and use of mirror background also add much to brightening up the display. Glass shelving allows the available light to pass through to the next shelf, thus, reducing the shading effect of solid shelving. There is something about objects being displayed on lighted glass shelves that just makes the items more desirable. Have you ever been attracted to a nice looking piece of merchandise in an attractive store display case? The more you looked at it, the more you just had to have it. Later, at home, it just didn't seem to look as nice as it did in the store display case.

Security

Collecting farm models is expensive and many dollars can be invested over a period of time. Often a beginning collector eagerly is looking for and adding to his collection with all his attention to learning about what's available and where to find it. After a period of time, he begins to settle down and suddenly realizes there are a lot of dollars setting on those shelves. As with an other investment, such as his auto or home, he begins to become concerned with security.

There are several aspects to security. One that may concern many collectors is the unthoughtful visitor or fellow collector that just has the impulse to pick up and handle the various models. This can be very stressful to the owner. The dropping of a model could greatly reduce its value by chipping the paint or breaking it. A non-collector type visitor may not realize the dollar value of a particular model, thinking it to be just a "toy".

Children, either the collectors or those of visitors, can also cause concern. It's only natural for children to want to play with the models.

Security in these situations may mean the use of glassed in type display cases that make for uneasy access to your collection. Displaying on shelving above the reach of smaller children would be advisable. In some cases, a lockable door to the display room would keep unwanted visitors out.

Another security concern may be the theft of all or part of your collection. It is difficult to be 100% protected. A lockable room is recommended. Some collectors may even choose electronic sensor and alarms to assist in securing their collection. A good relationship with a near-by neighbor can be very helpful by encouraging them to feel free to investigate any unusual visitors while you are away from home, especially if gone over a weekend or on a vacation. Placing an identification code on each model may make it easier for law officers to track down and recover stolen items.

Lastly, another security concern will likely be that of insurance. When the collector begins to realize that he may have several thousand dollars invested in his collection, he may realize the potential financial loss that could occur from theft, fire, tornado, flooding, etc. As farm toy collection grows and as more publicity is given to collectors and their "high priced" toys, the probability of someone breaking in and stealing it is likely to increase. A number of collections have been stolen from in recent years.

Even though one's collection can never be replaced, it is some conciliation to be able to recover a monatary value from the loss. Nostalgic values, memories and replacement of those items in the collection that "belonged to grandfather when he was a boy", just can't be replaced.

More and more collectors are insuring their hobby. Some homeowner policies automatically include collectables along with other household items up to a stated dollar amount. This may be satisfactory for a smaller collection. Most insurance companies will require an itemized listing of all the items to be included in the policy with a stated value. They may request an appraisal by one or more "authorities" substanciating the realistic values of the items in your collection. Pictures of the collection many times will be required for record purposes. Annual updating often times is required.

Types of Displays

Many types of displays can be used to effectively show off your collection and also provide differing degrees of security.

The purchase of "used" store display cases provides a rather quick method to house ones' collection. The older oak and glass cases offer a flavor of "antique" to housing toys, especially the older vintage ones. The newer aluminum and glass type with indirect lighting make very attractive, dust free displays.

Motorized fast merchandising cases such as "Timex" and "Wesclock" make for relative inexpensive, neat displays. Often they are of the upright design, therefore providing good shelf space in a relative small floor area. The carousel feature allows for easy viewing from one side.

Wall type shelving is the most common display method. The use of wall brackets with adjustable shelf brackets are very common. They are readily available and easily installed. Being able to adjust shelf spacing can be a valuable feature especially as the collection grows and changes.

Solid shelving of the built in type is another choice of some collectors. Substituting glass shelves assists in reflective lighting and adds brightness to the models. The use of mirrors as a background gives a 3-D viewing effect. It also allows for the back side details to be seen without disturbing the model. The reflectiveness of the mirrors also adds to the illumination of the display. The addition of front cover glass provides an added measure of security and a more dust free condition.

Other display areas include: unused window casing converted into a recessed display, ledge over doorways in older type houses for display of 1/64 to 1/43 scale models, planks separated by brick or cement blocks to provide shelving, corner cabinets, shadow boxes, fireplace mantels, top of file cabinets, old book cases, etc.

The following selection of pictures illustrate methods of displaying the farm toy collection.

A machinery salesman's office display.

A smaller wall display suitable for home or office.

Home made, glassed in case handy for featuring a selected group of models.

An easy-to-construct box type shelf display featuring narrow shelves for displaying 1/64 to 1/43 scale models. Many models can be exhibited in a small space.

10

Older, oak and glass store display cases used for dust-free type of exhibit.

Unused window converted into a display.

Example of an out-building re-furbished with an "antique" flare. Barn siding for wall cover, plank floor to give the warmth of older days.

Built-in units with indirect lighting, glass shelves and storage cabinet at base. Effective utilization of a room corner also shown. Box display allowed at top of units. Sliding glass doors reduce dust accumulation.

Vertical glass display featuring storage cabinet at base, box display at top, round glass shelves supported in center with lighting in top of unit. When placed in center of room, models can be viewed from four sides.

This view shows corner display usage adjacent to the two wall displays. Vertical glassed in cases can also be seen in this picture.

Unitized, built-in display, allowing for displaying collection by groups. Note adjustable shelf brackets. This unit extends all the way to the floor. Larger models display better on bottom shelf.

A light background coupled with the movement of light through glass shelves gives models a brighter appearance.

A combination open shelf display, with glassed in case type display area beneath.

This arrangement allows for storage under display shelving, glassed in, lighted display of models on glass shelves and box display over head. A clean, neat, secure display.

"Boxes"! They Tell a Story

Several years ago, this writer attended a farm auction, in which, a toy model Cockshutt tractor was advertised. Since I have been diligently seeking one of these steerable Cockshutt 30 models for sometime, anxiety was building up as the sale day approached.

When my wife and I arrived at the auction site, we hurriedly began to look for the advertised model. Suddenly among the "junk" wagon, I spotted a box showing the image of a "Cockshutt" toy tractor. My heart began to pound as I worked my way to inspect it more closely. Upon opening the box, I reached in to pull out the toy tractor. My heart sank to discover that it contained a 4010 John Deere painted red by its former young owner. Having had the rug pulled out from under us, we debated what to do now—stay for the auction or begin the long drive home. Knowing that the box existed and that at one time a Cockshutt tractor came in it, we began to locate the owners—perhaps they might know where the tractor was. Again, we were confronted with bitter disappointment.

Going back home empty handed is not my idea of a feeling of accomplishment, especially after spending the good part of the day and many miles of travel.

Why not buy the red tractor for the box? After all, it may be the closest I ever come to having a Cockshutt 30 in my collection.

Finally the boxed tractor is brought up to the auction block. The adrenaline begins to flow, my heart is rapidly pounding as the auctioneer removes the "Red" tractor from the Cockshutt box and begins his "speel." Somewhat apprehensive to bid—a grown adult bidding on a kid's toy—I threw in my bids. "Sold!" I had won the bid. With a sigh of relief, I worked my way up to pick up my prized purchase—a Cockshutt box. Upon reaching the auctioneer, he handed me the "Red" tractor. "Where's the box?" I nervously inquired. "I threw it under the table." he replied. "All I wanted was the box." was my reply.

While the crowd's eyes fixed on this "dummy" that paid that amount just for the box, the auctioneer's helper fetched the discarded box for the "kid."

Happy as a child on Christmas morning, we headed back home with my prized posession—a Cockshutt 30 box.

As collecting farm toys has grown and matured, interest has increased among some collectors in having original boxes to compliment their collections, especially of the older models.

"Mint in the box" has often been considered the ultimate in collection excellence. A growing number of collectors are enhancing their display of models with their original boxes. It is becoming a more common event to see empty boxes offered for sale on dealer tables at toy shows and auction blocks.

Why the interest in "boxes?"

Aside from the fact that finding an older or rare model in its original box as being the "status" symbol of the elite collector, there are other purposes of including boxes in ones collection.

Farm toy packages contain various kinds of information. Some of the informative details may include:

1. Scale: 1/16, 1/32, 1/20, etc.
2. Name and location of toy manufacturer.
3. Type of construction material.
4. Name of the toy—often a "promotional" of a real machine.
5. Description of possible "working" features of toy.
6. Picture, drawing, or illustration of enclosed model.
7. Country manufactured in.
8. "Stock" or "Article" number.
9. "Logos" or trademarks the company the toy represents.
10. On occassion, brief histories or promotional information is given of the "real" machine represented by the toy model.
11. More recently, "child age" recommendation.

Boxes and packages often are used to "authenticate" the originality of a model. They may also serve as the "authoritive" information source for details.

Purpose of Packaging

The primary purpose of packaging toys and models is to protect them from damage during shipment.

In early days of manufacturing, toys were usually shipped in bulk. Twelve, twenty-four, or forty-eight items, as an example, would be shipped in one carton or crate. Some type of packing material—old newspapers—would be used to prevent damage within the container.

Later on, individual boxes were used to put each item in. Earlier boxes were known as "chip board" type of cardboard material, often void of any markings or information with only the name of the toy stamped on the box. Following this, various types of information was printed on the box.

During the 1950's era, manufacturers began to dress-up the boxes to make them eye appealing. An example of this is the green and yellow chipboard boxes used to package the high post John Deere Tractor and various implement of that time. Color added much to the attractiveness of the boxes, along with the illustration of the model displayed on the box.

As time went on, packaging technology progressed greatly through out all of industry, including toy manufacturing.

A secondary purpose of packaging other than protection began to receive attention—that being eye appealing merchandising. Beginning about the early 60's, the era of the 4010 to 5020 type John Deere models, the white cardboard boxes with four color printing began to adorn dealerships catching the attention of the "young farmers."

Tugging at dad's shirt tail often resulted in some young boy going home with a new tractor just like dad's real one.

Discount merchandising outlets also began to sell many of the farm toys, which at one time were only available through dealerships of farm machinery. The need for attractive display packages greatly enhanced the sale of farm toys through mass merchandising outlets.

The use of "see through window" packages along with "blister" packages were often examples of the "eye catching" types of packaging for mass merchandising.

Who Decides on Type of Packaging?

Many factors as considered when selecting the type of package that the toys are to be marketed in. Often, both toy manufacturer licensee and the company they make the toy models for, "licenser", work together in packaging design.

Some of the considerations include:

1. Ease of placing items in packages during production.
2. Protection of model.
3. Point of purchase attractiveness.

4. Desired kind of promotional information on package.
5. Cost of package.

Much attention is given to the difficulty of securing the toy into its package during the final stage of production. The more time spent on this phase of production, the more it adds to the costs of the item; therefore, various means are used to quickly and effectively secure each model into its temporary home. The Britains Company of England used a "lock-in'key" to hold the model to its package. Many U.S.A. models are packaged in corrigated cardboard boxes that have been die punched to provide locking tabs, wheel recesses, and/or axle down tabs to anchor the model.

When providing protection, in some instances, the type of item dictates the kind of package. Plastic kits require a "box" type package to reduce the possibility of fragile parts from being crushed.

If the farm toy model is a blue-print promotional type, often the manufacturer of the real machine is interested in relaying some specific information or company image. This is accomplished by utilizing a multi-colored lithographed picture of the real machine along with a story of the features of the real machine the model represents. The Massey Ferguson 2680 plastic kit by "Heller" of France in their "Bobcat" Series is an illustration.

A few years ago when the White Motor Company sold its Farm Machinery Division to another company, the new owner was very interested in conveying the new company image to the public. As a result, the old skematics used on the earlier "White" toy box was dropped and replaced with the new company logo—a new identity.

Packaging adds to the purchase price of the toy; therefore, costs are a factor in the final decision. Approximately 5 percent to 10 percent of the toy's cost may be in the form of packaging. This cost is even a higher percent on small, low cost toys such as a $2 blister pack item.

Adding colorful, informative boxes to one's collection can add to the total satisfaction of "completeness" to one's hobby. Some collectors have been known to receive nearly as much "kick" out of adding a nice box to their collection as they receive from the model itself. To each his own.

Example of early type boxes. Plain cardboard with rubber stamp printing listing contents.

14

Boxes of the 1950s were highly detailed, showing pictures or illustrations of contents, trade marks, manufacturer, etc. Color added much to the eye-catching boxes.

Sample of cellophane covered box used to discourage children from playing with the model or removing the rubber muffler.

Selection of boxes used for packaging models in "kit" form.

Special packaging for large catalog sales companies were boxed similar to this "Wards" package.

Illustrations of packaging used by several non-U.S.A. countries.

VICKERS VIGOR

REMOTE CONTROLLED
TRACK LAYING

TRACTOR

TRACK STEERING

POWERED BY
TWO INDEPENDENTLY REMOTE CONTROLLED MIGHTY MIDGET
ELECTRIC MOTORS

ANOTHER VI Product

MODELLED BY
VICTORY INDUSTRIES (SURREY) LTD., GUILDFORD, ENG.

IN COLLABORATION WITH
VICKERS-ARMSTRONGS (TRACTORS) LTD., NEWCASTLE-UPON-TYNE, ENG.

MADE IN GREAT BRITAIN

AUTHENTIC SCALED MODELS OF
MASSEY-FERGUSON
FARM EQUIPMENT

MF

MADE BY
LIT'L TOY

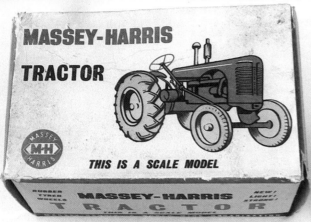

MASSEY-HARRIS
TRACTOR

M-H

THIS IS A SCALE MODEL

RUBBER TYRED WHEELS   MASSEY-HARRIS   NEW! LIGHT! STRONG!
TRACTOR

CASE

PLASTIC
TRACTOR

Case-a-matic DRIVE TRACTOR

MINIATURE MODEL CASE 800 TRACTOR

COCKSHUTT MINIATURE
TRACTOR
RUBBER TIRED · ALL METAL

Made by ADVANCE PRODUCTS CLEVELAND, OHIO

New Idea
Junior-Farmer
tractor mower

more than a toy
. . . a real scale model!

remember . . . if it's a
New Idea
it's a good idea

New Idea
Junior-Farmer
corn picker

OLIVER ROW CROP
77
TRACTOR
A SCALE MODEL
ALUMINUM TOY

OLIVER
MECHANIZED
MOWER

The OLIVER TRACTOR-MOUNTED
CORN
PICKER

Easily mounted on
OLIVER "77" Toy Tractor

MINNEAPOLIS-MOLINE
Hi-Klearance
PLOW

A Sturdy Scale Model
FARM
TRACTOR
IT REALLY STEERS!

MINNEAPOLIS-MOLINE
445
POWER-LINED
TRACTOR

Realistic Model Tractor
MINNEAPOLIS-MOLINE
MM

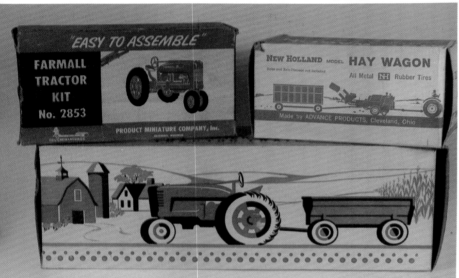

A selection of foreign and domestic boxes. Bright colors, illustrated or pictured model on the package and interesting facts and information add authenticity to the model when added to the collection.

060518-008 (TE-30) 1/20 AIRFIX

061719-054 (F) 8'' BING

061719-055 (F)/HALF TRACKS 8'' BING

160518-002 7300 1/16 ARNOLD

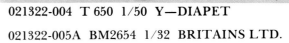

021322-004 T 650 1/50 Y—DIAPET

021322-005A BM2654 1/32 BRITAINS LTD.

130502-009 MB TRAC 1/87 BRUDER

131300-004A (R) 1/16 AUBURN RUBBER

131300-004B (R) 1/16 AUBURN RUBBER

19

180514-004A, -004B R30-40 1/16 CIJ

060514-005 FAVORIT 4S TURBOMATIK 1/43
CURSOR, 060514-010 FAVORIT 620LS TURBO
1/32 GAMA, 060514-001 ?? 1/25 CURSOR

100400-005 FROELICH, -006B WATERLOO
BOY (R) 1/16 CHARLES COX

081512-001 CULTITRAC A55 1/30 CURSOR
081512-002 CULTITRAC A60 1/27.5 CURSOR

040521-025 D-40L 1/21 CURSOR

060514-022A F275GT, -003 F250GT, -022B F250GT 1/43
CURSOR

061301-001A SERIES I 1/43 DINKY

120525-001B, -001A 384 1/43 DINKY

112102-006 M79500T 1/20 DIAPET

190113-002 CENTAURO DT 1/15 DUGU

010300-017A B-110, -018 LGT 1/16 ERTL

010300-019B 190, -023A 200 1/16 ERTL

010300-024D, -032B 7060 1/16 ERTL

010300-044A, -044B 4W-305 1/32 ERTL

030119-007B 1070AK, -008D (1370) AGRI-KING 1/16 ERTL

030119-007C 1070 AK DEMONSTRATOR, -007D 1070 AK451 DEMONSTRATOR 1/16 ERTL

030119-008F, -008G SPIRIT OF 76 1/16 ERTL

030119-005B 1030 1/16 ERTL, -003B (SC) 1/16 MONARCH PLASTIC

061718-026A 4000, -026B 4400 INDUSTRIAL
1/12 ERTL

030119-052B 3294, -025B 2590 1/16 ERTL

090803-030D, -031B CUB CADET 129 & 1650
1/16 ERTL

031503-005A, -005B 1850 1/16 ERTL

030119-053 2294, -038A 2290 1/32 ERTL

090803-039A 544, -039F (2644) INDUSTRIAL 1/16 ERTL

100400-023B (B) 1/16 ERTL

090803-091B CUB CADET 682, -091E CUB CADET (682)
1/16 ERTL

100400-026B, D, C, E 140 1/16 ERTL

090803-100A 5288 1/16 ERTL

100400-030A (630) 1/16 ERTL

100400-030C (630) 1/16 ERTL

130106-034B 1155 1/16 ERTL

100400-042B (40) INDUSTRIAL CRAWLER, -042A (40) CRAWLER 1/16 ERTL

151209-017B 1800 (SERIES B), -017G 1850 1/16 ERTL

130106-029A 175 DIESEL, -029B 3165 1/16 ERTL

131300-013 (ROW CROP LPG), -012A (602LPG) 1/25 ERTL

131300-016 G-1355 1/16 ERTL, -005 (UB) 1/16 SLIK

120113-001A R1056 1/43 FORMA PLAST

192005-003B COUGAR III ST 250, -003D PANTHER ST
310 1/32 ERTL

040521-020 D10006 GAMA

192005-003A PANTHER, -003C PANTHER ST 310, -003E
INDUSTRIAL 1/32, 061718-072A FW-60 1/64 ERTL

040521-008 D10006A 1/29 GAMA

040521-020 D1006 GAMA

130114-001 ?? 1/32 HERBART

040521-019 FIL514 1/20 HAUSSER

061718-016B 4000, -016A (4000) 1/12 HUBLEY

130106-060 2680 1/24 HELLER-BOBCAT (HUMBROL)

061718-015A 961 POWERMASTER 1/12 HUBLEY

020111-001A 21-75 SPECIAL 1/25 IRVIN

100400-004 (D) 5'' KANSAS TOY

220905-001 401 1/30 JRD

260520-009 CRYSTAL 8011 1/18 KDN

031503-006 30 1/16 KEMP PRODUCTS LTD.

010300-026A, -026B HD-15 4'' LIONEL

090803-014A (M), -014C (M)/STARS 1/16 PRODUCT MINIATURE

120126-003 BULLDOG 1/43 MARKLIN

090803-014E M 1/16 PRODUCT MINIATURE

051903-002, -003 335 1/20 MAXWELL

100400-104A, -104B 3185 1/43 NPS

180514-007 651 or 652 1/13 MONT BLANC

090803-153 SUPER C 1/16 PIONEER TRACTOR WORKS

160719-001 420 ROMA 1/9 PROTAR

220518-003A, -003B 1150 1/32 SCALE MODELS

142106-002 (UNIVERSAL) 1/15 RAPHAEL LIPKIN, -001
1/16 DENZIL SKINNER

030118-001B 88.4, -001C JUMBO (88.4) 1/24 SCAME-
GIODI

130108-010A (745), -010B 745 1/20 RAPHAEL LIPKIN

230809-001A, -001G 2-135 1/16 SCALE MODELS

191513-003 40 1/20 SEVITA

120114-003 R5000N SPECIAL 1/15 SIACA

100400-161 730 1/16 SIGOMEC

100400-163 ?? 1/16 SIGOMEC

100400-063A, -063B (4230) 1/16 SIGOMEC

131300-008 445 1/32 SLIK TOYS

100400-062A, -062B (4230) 1/64 SIGOMEC

131300-006 (R) 1/32 SLIK TOYS

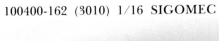

100400-162 (3010) 1/16 SIGOMEC

151209-010 880, -011A OC-6 1/16 SLIK TOYS

180514-008A, -008B 651-4 1/32 SOLIDO

020907-001 360/30 1/16 TRUMM-DEBAILLIE

020412-002 (MTZ-62) 1/43 SOLIDO-U.S.S.R.

090803-173 C 1/16 TUDOR ROSE

060518-018 TE-20 1/20 TRITON (LEGO)

120114-002A 1/43 YAXON, -001A R-4000 1/43
MERCURY, -002A 1/43 YAAXON

192018-001B 8160b, -001A 8160a 1/43 YAXON

211819-002 C335 1/20 UNKNOWN

040521-005A, -005B 06 SERIES 1/32 ZISS R.W. MODELLE

190114-001 BEAVER 1/32 UNKNOWN

012220-001 (MTZ) 1/16 MINILUXE, 020412-001 420 (MTZ 52) 1/16 MINILUXE

130106-038 165 1/16 UNKNOWN

□PT0103-006B　　　"190 XT"　　　ERTL　　　37" long

□PT0301-001　　　(VAC)　　　ESKA　　　35" long

□PT0103-008　　　"7080"　　　ERTL　　　37" long

□PT0301-002　　　(400)　　　ESKA　　　38.5" long

□PT0103-010　　　"7080"　　　ERTL　　　37" long

□PT0301-003　　　"CASE-O-MATIC"　　　ERTL　　　38" long

□PT0301-007 "CASE" (90 SERIES)   ERTL 35.5" long

□PT0615-004   "8000"   ERTL   36" long

□PT0615-001 (900)   GRAPHIC REPRODUCTION 40"
long

□PT0615-005   "TW-20"   ERTL   36" long

□PT0615-009   ?   Falk   110 cm long (including trailer)
Plastic, tri-cycle front. Available with two wheel trailer.

□PT0615-003 "COMMANDER 6000"   ERTL 38.5" long

□**PT0615-010** "TW-10" **Falk** 27" with dump box
Plastic, tri-cycle front, equipped with dumping box on rear.

□**PT0405-001B** "5006" **ROLLY TOYS** 35" long

□**PT0615-011** "TW-30" **Falk** 90 cm long
Plastic to -008 except decal designation.

□**PT0405-002** "4.70" **ROLLY TOYS** 31" long

□**PT0405-001A** "5006" **ROLLY TOYS** 35" long

□**PT0405-001B** "5006" **ROLLY TOYS** 35" long

□**PT0405-004b** **"DX 110"** **ROLLY TOY** **128 cm**
including loader
Same as -004a except front end loader.

□**PT0605-001** Favorit Turbomatik **"Rolly Toys"** **39.5"**
long
Plastic.

□**PT0405-005** **"DX85"** **FALK** **22.5" long**

□**PT06090-001** **"880"** **Fuchs**
Orange, plastic, displaying typical Fiat markings.

□**PT0405-006** **"DX 160"** **Falk** **29.5" long**
Plastic, with horn and cleated front drive wheels.

□**PT0908-005** **"400"** **ESKA** **39" long**

□PT0908-006   "450"   Eska   39" long
Smooth grill pattern with decal for design. Other features same as -005.

□PT0908-007   "560"   Eska   38" long
A casting change representing the new six cylinder series of Farmalls replacing the former four cylinder models. Decals used to accentuate detail. Finger grip steering wheel, metal end tear drop pedals, pressed steel seat, spoke pattern wheels.

□ PT0908-008　　　"806"　　ERTL　　36.5" long

□ PT0908-011　　(66 Series)　　Ertl　　37" long
Model designation no longer used. Decals are self-adhesive type. Plastic front wheels using push nuts for fasteners. No hub caps. New casting design having code 404.

□ PT0908-009　　　"856"　　Ertl　　36.5" long
This model represents the name change from "Farmall" to "International". Same casting as -008.

□ PT0908-012　　(86 Series)　　Ertl　　37" long
Same casting as -011. Plastic front wheels, fat plastic steering wheel begun being used with this model. Black strip hood decal.

□ PT0908-010　　"1026 Hydro"　　Ertl　　36 .5" long
Same casting as -008 and -009. Casting code I-64.

□ PT0908-013　　　"844"　　Rolly Toys　　?
Plastic, European utility type model. Detailed grill design.

□**PT0908-014** "844" **Rolly Toys** 65 cm long
Plastic, single front wheel pedal drive model.

(126) **250,00**

□**PT0908-016** "1055" **Falk** 91 cm long
Plastic, European styled model with nicely detailed three dimension grill. Lug type front and rear tires.

□**PT0908-017** "633" **ROLLY TOYS** 24.5" long

□**PT0908-018** "1455XL" **Falk** 36" long
Similar styling to -016.

406AO

□**PT0908-020** "International" **Ertl** 36" long
This model introduces a new construction from Ertl. Its' first plastic constructed pedal tractor. Styling features the new 50 series International tractors. This model has a wide front axle, chain drive and knob type tread on rear wheels.

403AO

□**PT0908-021** International 7488 **Ertl** 21" long
A new twist from Ertl. Not a pedal tractor, but a "ride-on" type model. Made of break resistant plastic. Cab door opens for storage. Exhaust stack doubles for steering handle.

□PT1004-001   (A)   ESKA   33.5" long

□PT1004-009   (4430)   ERTL   37" long

□PT1004-002   (60)   ESKA   34" long

□PT1004-008   "LGT"   Ertl   35.5" long

An unusual pedal tractor model in that it represents a "Lawn and Garden tractor". Has wide front axle, formed plastic grill insert, smooth rear tires, and pressed steel seat and fender covers. Casting code DTG-70.

□PT1004-003   (60)   ESKA   38" long

□PT1004-010   (4440)   Ertl   37" long

Same as -009 except "strobe" design on decal corresponding with the real tractor.

□**PT1004-011** (4450) **Ertl** 37" **long**
Same casting as -009 and -010 except decal simulates the
series of lights in front of grill as used on the real tractor.

□**PT1004-012B** "3140" **Rolly Toys** 128 cm **long**

□**PT1004-012b** "3140" **Rolly Toys** 128 cm **long**
including loader
Same model as -012a except equipped with front end loader.

□PT1306-003A "188" Rolly Toys 30.5" long
Plastic model of European styled M.F. tractor.

□PT1306-004b (188) Rolly Toys 128 cm long
including loader
Same as -004a except equipped with front end loader.

□PT1306-002 (2775) ERTL 36.5" long

□PT1306-005 "145" FALK 36" long

□PT1306-001 (1130) ERTL 36.5" long

□PT1306-006 "100" Falk ?
Plastic, tri-cycle, single front wheel style. Blue seat, yellow
steering wheel and fenders.

□**PT1306-007**   **"2640"**   **FALK**   30" long

□**PT1309-001**   **"Super S-150"**   **Fuchs**   ?
Plastic, non-promotional casting showing similar lines to
Fiat 880 by Fuchs (PT0609-001).

□**PT1309-002**   ?   **FERBEDO**   96 cm long
Plastic, non-descript pedal tractor featuring blinking lights.
Light green with silver engine and grill, black fenders and
seat. Front tires feature cleated lug design.

□**PT1308-002**   **"44 SPECIAL"**   **ESKA**   38.5" long

□**PT1302-001**   **MB Trac 1000**   **Falk**   75 cm
A promotional model of the Mercedes-Benz tractor displaying the MB trademark in front grill. Chain drive with cleated
front wheels. Light and dark green color combination.

285 Rolly Traktor
92 x 43 x 60 cm, 1 Stück, 195 cdm, 9,0 kg

□PT1309-003   "285"   Rolly Toys   92 cm long
Plastic non-descript model with Rolly Toys name in upper part of grill. Has cleated front drive tires.

484 Rolly Frontlader
124 x 43 x 60 cm, 1 Stück, 195 cdm, 11,0 kg

□PT1309-004 "Rolly-Traktor"   Rolly Toys 124 cm long
Plastic blue and red trim, non-descript tractor equipped with front end loader.

□PT1313-001   "TOT-TRACTOR"   BMC   40" long

□PT1309-008   "BIG"   ?   ?
Plastic, tri-cycle front end with drive pedals. "BIG" embossed on hood. Red seat and wheels, white steering wheel. Made in West Germany.

□PT1512-001 "88 DIESEL POWER"   ESKA   38" long

45

☐ PT1512-004   "880 DIESEL"   ESKA   39" long

☐ PT1805-002 "TURBO TX 133-14"   FALK 29.5" long

334. Tracteur à remorque.
Tracteur à pédales avec sa remorque. Roue AV directrice antidérapante. L. 110 cm. . . . . F 164,00

☐ PT1805-003   "RENAULT"   FALK   110 cm long

< ☐ PT1512-006   "1800 C"   Ertl   39.5" long
Cast aluminum. Stock #1004. Red border side decals, plastic grill, casting code 0-63.

☐ PT1805-001   "1451-4"   AmpaFrance   33" long
Plastic, Renault orange with white fenders. Front cleated tires and vertical muffler.

331. Tracteur Renault.
Vitesse 4 km/h. Batterie 6 volts livrée avec rechargeur sur secteur. Démarrage au pied.
86 x 41 x 51 cm. . . . . F 900,00

☐ PT1805-004   "RENAULT TURBO"
AMPAFRANCE   86 cm long

□PT1805-005   "951-4"   **FALK**   36" long

□PT1805-006   "751-4"   **FALK**   29" long

"A home town parade float displaying pedal tractors".

☐PT2308-001   "C-195"   **ROADMASTER CORP.**
39.5" long

Author and wife with a few pedal tractors from the collection of Bill Barlett Jr. of Illinois.

Farm Toy Advertisements

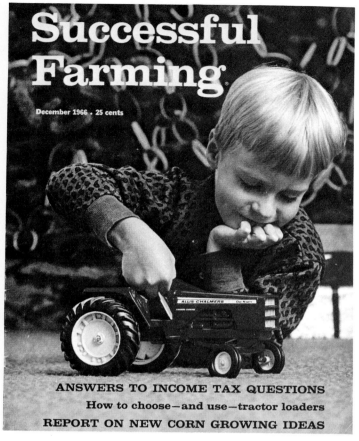

Allis Chalmers used on the December 1966 cover of SUC-CESSFUL FARMING.

I.H. used on the January-February 1984 cover of FARMING.

CUB TRACTOR from POPULAR SCIENCE, November 1947.

"JUST LIKE DAD'S" in December 1954 COUNTRY GENTLEMAN.

Cub Tractor —Toy Version

WHILE production was under way for delivery of the first new Farmall cub tractors to America's farmers, the International Harvester Co. also was tooling up on a pint-size toy version for Junior.

The model, shown here being assembled, is a faithful reproduction of the McCormick-Deering Farmall, scaled down to 1/16 size. It is six inches long by three inches high and can be assembled and knocked down at will. Clips are used for fastenings.

The tractor's interlocking parts are made of Tenite #2, a butyrate plastic. This material was used because, while medium-hard and durable, it is both pliable and flexible. Its natural luster also shows off the familiar Harvester equipment red.

Tires of the small tractor are of real rubber, and on later models they will have treads. Although this first production model of the toy is not mechanized, plans are under way to include this feature in improved versions. Also being developed is a line of quick-change farm implements that can be assembled and attached just like those on the big tractor Papa operates.

"Just Like Dad's!" By M. Estle Brown and Merle Betts

Solve the toy problem with working models of your favorite brand of farm machinery.

Country Gentleman December 1954

REALISM in toys is the trend today. And what can be more realistic to a farm boy than rugged, working models of the same equipment his dad uses. These toys are identical with their life-size counterparts right down to the maker's name and color. They include tractors, combines, manure spreaders, wagons, plows, rakes, harrows, balers and mowers. Many can be taken apart and reassembled over and over again. It's even possible to get repair parts. Some of the toys are plastic and some metal. Prices range from 60 cents for a plastic wagon to $3.75 for a metal combine with working parts. In most cases, you'll find these toys available from your regular farm-equipment dealer.

Junior feels he's doing his share when he pedals a riding-size tractor like this around the place. For ages three to eight. Price range is $20 to $30.

TWO HAPPY BOYS WITH DEE BROTHERS TOYS
Where One Goes, the Other Follows

Dee Brothers Farm Toys
New 1946 Models
Farm Toy Tractor and Wagon

Nothing Like Them On The Market Today   Test Sales Above All Expectations

---

SPECIFICATIONS

TRACTOR   Selected Quality Wood - Metal Type Finish
      Length 11¼" - Height 5½"
      Finish—Tractor Red, Black Tires and Trim

WAGON   Length 11¾" - Height 6⅜"
      Box—Wagon Box Green—Demountable
      Running Gears—Red with black treaded tires.

HARDWOOD AXLE BEARINGS ON BOTH MACHINES

Built in our own factory—each item is carefully tested and inspected. Sold through established implement dealers. Initial orders limited to one dozen each.

DEE BROTHERS FARM TOYS

D. & D. PRODUCTS   FORT DODGE, IOWA

FARMALL TRACTOR & WAGON—DEE BROS.

EXCLUSIVE JOHN DEERE FARM TOYS by ERTL

OH BOY! JUST LIKE DAD'S

Toy Model Cockshutt "30" Tractor and Wagon Set

The realistic appearance of this toy tractor set will catch the eye of every "young farmer". The set is a replica model of the popular Cockshutt "30" tractor and "95" wagon and is sturdily made of heavy plastic. Both tractor and wagon have swivel mounted frame axles and are equipped with rubber tires. (Tractor or wagon can be purchased separately.)

COCKSHUTT tractor and wagon on Cockshutt Farm Equipment F39-3-53-55M.

MASSEY-HARRIS Combines & Tractors

They don't make
snowmobiles.
Or trucks. Or cars.

Just farm machinery.

They still pour their own iron, build a
good planter, and hand-polish their
plow bottoms.

My dealer knows me and my farm. Not just
my account number.

Some people wonder if a company with
these values can last. But their equipment
sure does. So I figure they'll still be
around when he's farming.

Maybe White Farm Equipment isn't for
everyone. But it's for me.

WHITE FARM EQUIPMENT

WFE ad in February 25, 1984 PENNSYLVANIA FARMER.

Steve, Dan and Andy Scharber, fourth and fifth generations at Scharber & Sons.

"ERTL toys have become a big part of our tractor business."

Steve Scharber, Scharber & Sons, Inc., Rogers, Minnesota

For year-round profits and traffic, ERTL toys are more than just a sideline at the John Deere dealership of Scharber & Sons.

Steve Scharber says, "Kids help sell big tractors because they know if their dad's going to buy a big one, we're going to give them a little one."

Customers buy ERTL toys year round for gifts. "A lot of wives come in and do their Christmas and birthday toy shopping, and the husbands usually come along and shop for something else." Scharber adds, "Where we used to sell the riding tractors mainly at Christmas, we now sell three to four times as many in summer." And one farmer bought Scharber's entire first shipment of ERTL's new radio controlled tractors for his nephews. "That's the toy for kids eight to 50," Scharber says.

Why do customers keep coming back for ERTL Blueprint Replica® toys? "Because of their reliability, the quality that goes into them," Scharber explains. "Some of those get used pretty hard and they hold up real well."

See how ERTL toys can help you build profits and traffic, too. Call or write ERTL today.

The ERTL toy: Copy of the real McCoy

ERTL
TOYS AND MODEL KITS
The ERTL Company, Dyersville, Iowa 52040
(319) 875-2481

Subsidiary of Kidde, Inc./In Canada: Kidde Recreation Products/Division of Kidde Canada Limited/543 Conestoga Blvd., Cambridge, Ontario Canada N1R 6T6

August 7, 1981

Circle (12) on Reply Card

21

Ertl ad in IMPLEMENT & TRACTOR, August 7, 1981.

DON'T
toy around
with farm safety

Play it for real
because you're playing for keeps

Equip your John Deere Tractor
with a new Roll-Gard

"Don't Toy Around With Farm Safety"

51

Scale Model of the

DAVID BROWN
Cropmaster 6

Farmers the World over from Norway to New Zealand will recognise this model at first glance as a faithful replica of the all-British "Cropmaster" tractor, one of the DAVID BROWN range of farm machines, which includes trailers and Power Controlled implements.

SCALE MODEL OF DAVID BROWN.

INSTRUCTIONS AND PARTS LIST
ASSEMBLY INSTRUCTIONS FOR DAVID BROWN 25 TRACTOR MODEL

To assemble the tractor it is suggested that it be put together in the following order :—

Chassis and Gear Box (D.31)

The Steering Column (D.11) will be found fitted with a Bevel (D.15) and will be assembled with Drop Arm (D.12) and Drop Arm Bevel (D.14) forming the Steering mechanism.

Two Gear Levers (D.43) are a push fit on the top of the Gear Box.

Two Battery Boxes are screwed on to the chassis each side.

Take Radiator Cowl (D.37) and screw into lug at front and on top of chassis with small screw. Place Fuel Tank (D.34) in position in slots in chassis and secure with long split pin through side of chassis. Take Engine (D.35) assembled with Fan and push into centre hole in chassis. Screw Bonnet (D.39) to Fuel Tank and top of Radiator Cowling with two more screws. Slide Engine Side Covers into position and screw Air Cleaner on to chassis with small screw.

Back Axle Assembly

Offer up Transmission Case (D.22) to two lugs on rear chassis. Push Axle Rod (D.23) through to secure. Thread R and L hand Mudguards (D.20, 21) into position over Axle Rod followed by Reduction Boxes (D.19) which when in position are screwed home by two 5BA nuts at the end of Axle Shaft. Assemble Seat (D.24) into position and secure with split pin. Assemble Pulley (D.25) on to power take-off with split pin. The two Rear Wheels (D.18) can be assembled at this stage or later.

Towing Gear

Take (D.26), Towing Attachment Bracket and assemble with swivel pin (D.27 to D.28) Tow Bar Guide Bracket. (D.30) Tow Bar will be attached to (D.28).

Underneath Transmission Case (D.22) which has previously been assembled to chassis will be found the position and holes for screwing complete Towing Attachment to the Transmission Case.

Front Axle

Select Front Axle (D.1) and with the longer section from centre hole held to the left insert R and L hand Stub Axles (D.5). Take

first, left hand Track Rod lever (D.8) and screw onto L.H. Stub Axle until approximately correct angle and secure with 5BA nut. Do the same to the R.H. Track Rod lever (D.7) and screw on to R.H. Stub Axle and secure with 5BA nut. The Track Rod (D.10) can then be linked into the holes on R and L hand Track Rod levers and secured with two 10BA nuts. The Stub Axles can now be correctly aligned by final adjustment on R and L hand Track Rod levers and nuts securing. Place end of Steering Rod (D.44) into hole on L.H. Track Rod lever and secure with 10BA nut. The loop end being put round lower end of Drop Arm and secured with 8BA nut.

Screw Steering Wheel (D.16) to Steering Column (D.11) first screwing on 8BA nut down to bottom of thread. When Steering Wheel is screwed on to correct height on column, tighten 8BA nut back on to Steering Wheel to lock. The whole steering mechanism can now be tested.

Finally assemble Front and Rear Wheels on to Front and Rear Axles, the convex part of the Hub facing outwards, insert Exhaust and make any final adjustments.

PARTS LIST FOR DAVID BROWN 25 TRACTOR MODEL

FRONT AXLE AND STEERING			
D.1 Front Axle	D.11 Steering Column	D.23 Shaft and Seat Support	D.39 Bonnet
D.3 Split Pin	D.15 Steering Column Bevel	D.6 Nuts	D.40 Exhaust
R.H. Stub Axle	D.16 Steering Wheel	D.24 Seat	D.41 R.H. Side Cover
D.5 L.H. Stub Axle	D.13 Steering Wheel Lock Nut	D.25 Pulley	D.42 L.H. Side Cover
D.6 5 BA Nuts	D.1 Front Tyre	D.4 Split Pin	D.43 Gear Lever
D.7 R.H. Track Rod Lever	D.2 Front Wheel	CHASSIS	
D.8 L.H. Track Rod Lever	BACK AXLE	D.31 Chassis and Gear Box	TOWING ATTACHMENT
D.9 10 BA Nuts	D.17 Back Tyre	D.32 Battery Box	D.26 Towing Attachment Bracket
D.10 Track Rod	D.18 Back Wheel	D.33 Air Cleaner	D.27 Pin
D.44 Steering Rod	D.19 Reduction Box	D.34 Fuel Tank	D.28 Tow Bar Guide Bracket
D.12 Drop Arm	D.20 L.H. Mudguard	D.35 Engine	D.29 Pin
D.13 8 BA Nut	D.21 R.H. Mudguard	D.36 Fan	D.30 Tow Bar
D.14 Drop Arm Bevel	D.22 Transmission Case	D.37 Radiator Cowling	
		D.38 Screws	

MANUFACTURED FOR DAVID BROWN TRACTORS (ENGINEERING) LTD.

BY DENZIL SKINNER & COMPANY LTD., CAMBERLEY

INSTRUCTIONS AND PARTS LIST—DENZIL SKINNER.

THE NUFFIELD "UNIVERSAL FOUR" TRACTOR

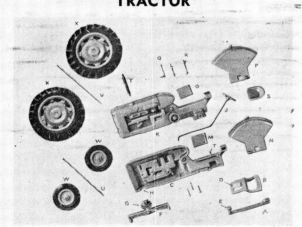

INSTRUCTIONS FOR TAKING APART

Pull off all Wheels from Axles, remove Mudguards and Footplates, take off Exhaust Stack and all control Levers, unscrew the three screws which you will see in the off-side half of the Tractor—the Tractor will then come apart and the other inside pieces can be removed.

INSTRUCTIONS FOR ASSEMBLING

1. Assemble Towing Bar "A" in its Quadrant "B" and lay the two pieces in position in the off-side half of the Body "C". (The rectangular lug "D" on the Quadrant will fit into the slot in the lower edge of the off-side half of the Body and the flanged pin "E" at the pivot end of the Towing Bar will drop into the half-hole notch in the bottom flange of the Body.)

2. Insert the Front Axle Housing "F" by locating flanged pivot "G" on Front Axle in half hole "H" on Body and the cranked end of pivot facing forward.

3. Locate Steering Wheel and Column "J" so that bend in Steering Column fits round central boss in Body and the cranked end fits into slotted hole in crank on Front Axle. (When properly located the boss of the Steering Wheel should fit snugly against the outside of the Body.)

4. Complete Body Assembly by fitting on near-side half of Body "K" and screwing in the three screws through the holes provided in off-side half section "C".

5. Fit very carefully Brake Lever "L" into hole provided on top side of off-side Footplate "M".

6. Fit Off-side Footplate "M" into position on Body with flange of Footplate downwards at rear.

7. Press on Off-side Mudguard "N" on to square lug provided on Body.

8. Assemble near-side Footplate "O" and near-side Mudguard "P" similarly.

9. Press near-side Lever "Q" into position in hole provided in body.

10. Press Gear Lever "R" into position in central hole provided on floor of Body.

11. Fit "Driver" Seat "S" on to pin "T" provided.

12. Insert front "U" and rear "V" Axles. (The rear Axle is the longer.)

13. Push on Front "WW" and rear "XX" Wheels.

14. Erect Exhaust Stack "Y."

RAPHAEL LIPKIN LIMITED

THE NUFFIELD "UNIVERSAL FOUR" TRACTOR, RAPHAEL LIPKIN LTD.

IT'S FUN ASSEMBLING YOUR ALLIS-CHALMERS SCALE MODEL "D" SERIES TRACTOR

ASSEMBLY INSTRUCTIONS

READ BEFORE STARTING ASSEMBLY

1 Study the instructions and become familiar with every part, (where it goes and how it fits with other parts) before you begin assembly. Follow the instructions step by step.

A very little "PLASTIC" Model Cement will be needed to glue this model together. Use cement sparingly, making sure surfaces are cemented are pressed firmly together. Many parts are locked in place without the use of cement.

2 **WHEEL PRE-ASSEMBLY**

1. Place rear tires over wheels, one with "R" up and one with "L" up. Push wheels firmly into tires.
2. Apply cement around outer flat areas of wheels and cement caps in place.
3. Cement two parts of each front wheel together. Allow fifteen minutes drying time before snapping tires onto wheels.

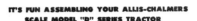

3 4. Insert shift lever and double foot pedal into holes provided in right body shell. No cement necessary.
5. Insert right foot platform projections into holes provided by working front of platform between the two arms of double foot pedal. Cement platform projections from inside body shell.
6. Slip single foot pedal onto front projection of left foot platform and secure platform in same manner as right side.
7. Insert air cap into left body shell and cement from inside.
8. Lay steering column into left body shell with key in proper place as shown.
9. Apply cement to edges and other meeting areas of left body shell and cement two shells together, thus locking steering column in place.

4 10. Snap tongue and side brackets of rear hitch into slots provided. No cement necessary.
11. Snap top projections of front axle unit into place as shown. No cement necessary.
12. Cement radiator and seat into place.
13. Insert rear axles into fender pieces. Rectangular portion of axles must fit flush into rectangular pockets of fender pieces. No cement necessary.
14. Insert rear axles and fender units into body, but before snapping them together check and make sure the "R" fender is on the right side and the "L" fender is on the left side. Arrows to the front, tail lights to the rear. Squeeze the axles together and listen for the click as they hook, one to the other. No cement necessary.

5 15. Snap rear wheels onto axles—"R" to the right side, "L" to the left side.
16. Snap front wheels onto axles. No cement necessary.
17. Snap steering wheel onto column, heavy portion of the hub down.

ALLIS-CHALMERS MANUFACTURING COMPANY
MILWAUKEE, WISCONSIN

ASSEMBLY INSTRUCTIONS SCALE MODEL "D" SERIES TRACTOR.

the NEW.

JUST LIKE DAD'S

Times and trends have really changed. . . instead of pestering their folks for the usual pony, the young farm-hands of today have made it known, loud and clear, that they want a tractor. . . just like dad's.

Well here it is at last. . . the Oliver Tractor-Cycle, a perfectly scaled reproduction of the popular Oliver "88" Diesel-Powered Tractor . . . the most enjoyable and educational toy any farm youngster could own.

Just put a Tractor-Cycle on display in your showroom and watch the young farmers that frequent your dealership with their dads get that "I want one look" when they try out these exciting features:

BUILT TO LAST . . .
Made of strong cast aluminum . . . can take plenty of rough treatment.

RUBBER TIRES . . .
Long Wearing . . . give a more comfortable ride. Cleated for traction . . . just like the real thing.

ADJUSTABLE SEAT . . .
Seat slides backwards and forward to fit child's leg length.

CHAIN DRIVEN . . .
Heavy-duty sprocket chain drives the rear ball-bearing wheels.

ACTUALLY STEERS . . .
Simple mechanism allows youngsters to steer without effort.

For further ordering information contact your Oliver Territory Manager

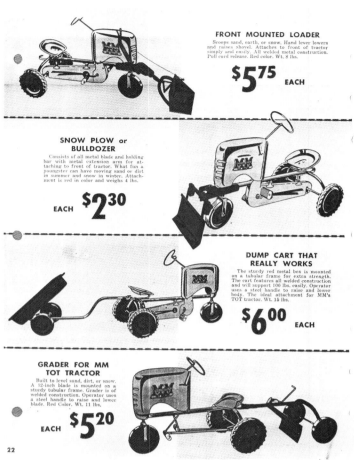
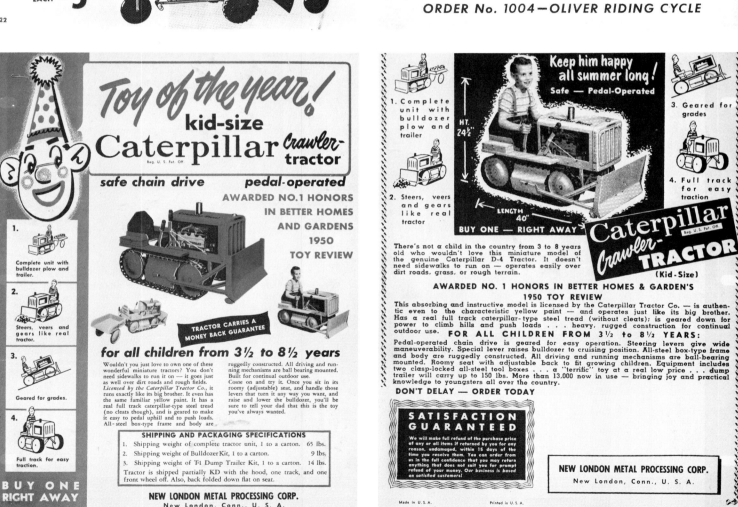

Model Farm Tractor and Construction Equipment Manufacturers 5

ABS Models—GB
A C Williams—USA
A H I—HK
A M T—USA
A R—F
A R France—F
A T & T Collectables—USA
Advanced Products—USA
Afinson—USA
Agfa Di Favero—I
Airfix—GB
Allemagne-Plasty—D
Alps Toys—HK
Aluminum Model Products
   (AMT)—USA
American Precision—USA
AMPA France—F
ANBMCS—GB
Anderson, Gary—USA
Animated Toy—USA
Apthyka—USSR
Arcade—USA
Arcor—USA
Arnold—D
Arpra Ltd—BR
Arthur Hammer—D
A T M A—BR
Aubrubber—USA
Auburn Rubber—USA
Auguplas—E
Auguplas-Minicars—E
Avon—USA

B M C—USA
B P—DK
Baker, Roy Lee—USA
Bandai—J
Banner Plastic—USA
Banthrico—USA
Barclay—USA
Barlux—I
Barr—USA
Benbros—GB
Ben Bros—GB
Bataltin Plate—D
Bijage Van Het—NL
Bing—D
Binghamton Mfg. Corp.—USA
Blue Box—HK
Bonux—F
Bourbon—F
Brio—S
Britains Ltd.—GB
Brown's Models—GB
Bruder—D
Buby—RA
Buddy—L—USA
Budgic—GB
Bukh—CO
Bull—D
Bulli—D
Burkholder, Charles—USA

C D—F
C I J—F
C I J Europarc—F
C K O—D
Cane Giogattoli—I

Carter Tru-Scale—USA
Cavallino—I
Chad Valley—GB
Champion—USA
Charbens—GB
Charmerz—HK
Cherrica-Phoenix (Taiseiya)—J
Chico Toys—CO
Clifford—HK
Clim—E
Cofalu—F
Co-Ma or Coma—I
Coma Intertoys—I
Conklin, Ken—USA
Conrad—D
Cordeg (Universal)—HK
Corgi—GB
Corgi (JR)—GB & HK
Cortland—USA
Cox, Charles—USA
Cragston—J
Crescent—GB
Cursor—D

D B G M—D
D C M T Lonestar—GB
D G Models—GB
Daisy—USA & CDN
Dee Bros—USA
Dent—USA
Denzil Skinner—GB
Design Fabricators—USA
Diapet (Yonezawa)—J
Dickie Spietzeug—D
Dingman, Lyle—USA
Dinky—GB
Dinky Dublo—GB
Doepke (Modell Toys)—USA
Dole—D
Dolecek—USA
Druge Bros—USA
Dugu—I
Dubena—CS
Duravit—RA

E. Allan Brooker—NZ
E R Roach—USA
Eberson, Alvin—USA
Eccles Bros—USA
Eidal Grip Zechin—J
Eko—E
Eligore—F
Empire—HK
Empire Made—HK
Enos Built—USA
Ertl—USA
Esci—I
Eska—USA
Espewe—DDN
Espewe Plasticart—DDN
Estetyca—POL
Ezra Brooks—USA

Falk—F
Farm-o-Craft—USA
Ferbedo—D
Fleetline—USA
Forma-Plast—I

Freiheit, Pete—USA
Frog—GB
Fuchs—D
Fun-Ho—NZ

G & S Frzeugnis—D
GBZ—?
G I H—D
G S—D
Gabryda—?
Galanite—S
Gama—D
Damda—IL
Gay Toys—USA
Gescha—D
Gianco—I
Gitanes—F
Goso—I
Gray, Robert—USA
Graphic Reproduction—USA
Grip (EIDAL)—J
Grisoni Giocattoli—I
Gubbels, Daniel—USA
Gusival—E

H.P. Plast (Nyrhinen)—SF
Hales—GB
Hallmark—USA
Hanson, Frank—USA
Hausser—D
Hausser-Elastolin—D
Heller Bobcat (Humbrol)—F
Herbart—D
Hiller—USA
Hoffman, Tom—USA
Hooker, Wally—USA
Hosch, Jim—USA
Hover—HK
Hoyuk—T
Hubley—USA
Hubley-Gabriel—USA
Husky—GB

Ideal—USA
Igra—CS
Ike Toys—DK
Imai—J
Impy (Lonestar)—GB
Inland Mfg. Co.—USA
Iron Art—USA
Irvin's Model Shop—USA
Irwin—USA
Ites—CS

J R D—F
Jadali—F
Jaya—E
Jean—?
Jergensen, Earl—USA
Joal—E
Johan—USA
Jo-Hillico—GB
Jordan Products—USA
Jouef—F
Jouef Mont Blanc—F
Joustra—F
Jue (Minimac)—BR

K—J
K & G Casting—USA
K D N—CS
Kansas Toy—USA
Karkura—D
Karran Products Ltd.—GB
Karslake, M. Howard—GB
Kaysun Plastics—USA
Keil Kraft—GB
Kelton—USA
Kilgore—USA
The King Company—USA
Kingsbury—USA
Kondor—?
Krishi—IN
Kurse, Marvin—USA
Kubota-Jigyosha—J
Kunstoff Herbert Fitzek—A

LBS—F
LBZ—D
Lakone—USA
Lakone Classic—USA
La Hotte-St Nicholas—F
Langley Miniatures—GB
Lansing—USA
Lee Toys—USA
Leningrad—USSR
Lesney (Matchbox)—GB
Lincoln Micro Models—NZ
Lincoln Specialities—CDN
Lincoln Toys—CDN
Lines Bros—GB
Lion Car—NL
Lion Molberg—NL
Lionel—USA
Lone Star—GB
Lucht-Friesen—USA
Lucky—HK

M I C—HK
M P C—USA
M S—China
M S—D
M S-Toy—D
M S-Michael Seidel—D
M/S—USSR
M T—J
M W—HK
Maasdam, Larry—USA
Machpi-Bologna—I
Machpi Caschi—I
Major Models—NZ
Majorette—F
Majorette-de Brazil—BR
Matchbox (Lesney)—GB
Marbil—GB
Marx—USA & HK
Mattell-Mini—IN
Maxwell—IN
Maxwell-Hemmens—GB
Meers & Son—GB
Mehand Teknica—JUG
Mehandicknica—JUG
Mercator—D
Mercury—I
Mercury Litl Toys—USA
Merit—GB

Mettoy—GB
Mettoy-Castoys—GB
Michael Seidel (M S)—D
Micro-Models—NZ
Midget Toys—F
Miltan—IN
Mini-Auto—D
Mini-Gama-D
Mini Toys—USA
Minic—GB
Minic-Triang—GB
Minimac (JUE)—BR
Mitoplast—BR
Mod-AC—USA
Modelle Toys (Doepke)—USA
MOdern Toy—USA
Moko-Lesney—GB
Monarch Plastic—USA
Moplas—I
Mont Blanc (Jouef)—F
Morgan Milton Ltd—IN
Multiple Products—?

N P S—HK
N Z G—D
Nacoral—E
Nasta—USA
New London Medal
    Products—USA
Noreda—?
Norev—F
Normatt—USA
North & Judd—S
Nostalig—?
Nostalgia—USA
Nova—F
Nygren, George—USA

OTT-PTW-Old Time
    Toys—USA Pioneer Trac-
    tor Works—USA
Old Cars—I
Old Time Collectables—USA
Old Time Toys—USA

P M I—ZA
P T W—Pioneer Tractor
    Works—USA
Pacesetter—USA
Parker, Dennis—USA
Parks, Brian—GB
Paya—E
Peetzy-Roco—A

Pending—?
Peter Mar—USA
Peterson's Model Shop—USA
Phanton Models—F
Piko—DDN
Pioneers of Power (Robert
    Gray)—USA
Pioneers of Power II—USA
Pioneer Tractor Works-
    (PTW)—USA
Pippin (Raphael Lipkin)—GB
Plasticum—?
Plasto—?
Plow Boy Toys—USA
Play Art Dickie—HK
Polistil—I
Politoys—I
Precision Craft Pewter—USA
Preiser—DDR
Processed Plastics
    Products—USA
Product Miniature Corp—USA
Promotion G. Belanger—CDN
Protar—I
Publi K—GB

Quelle (International)—HK
Quiralu—F

R.W. Modelle-Ziss—D
Raphael Lipkin (Pippin)—GB
Reindeer—ZA
Reliable—CDN
Renam—I
Replica Die Cast—AUS
Revell—USA & I
Rex—D
Roadmaster Corp.—USA
Roco—A
Rolly Toys—D
Ronson—USA
Rosko—?
Rowland Products—GB
Reuhl Products—USA

S G—D
S S S—J
Salco—GB
Sankyo—J
Sanson—?
Sanwa—J
Scale Craft—GB
Scale Models—USA

Scaledown—GB
Scame-Giodi—I
Schuco—D
Sevita—F
Shackleton-Chad Valley—GB
Sharna—GB
Sharp, David—USA
Sheppard Mfg. Co—USA
Siaca—I
Siegel, Ben—USA
Sigomec—RA
Siku—D
Slik—USA
Slik Toys—USA
Solido—F
Sommavilla—I
Souhrada, Charles—USA
Spot-On—IRL
Steam & Truck—D
Steho—D
Strenco—D
Strombecker—USA
Structo—USA
Sturdy Stuff Toys—USA
Stytrex—GB
Suede Pending
Sun Motor Co—GB
Sun Rubber Co—USA
Suttle—USA

T C O—D
T N—J
T T F—T
T Poli—I
Ta Tm chema
Taiseiya (Cherrica Phoenix)—J
Tekno—DK
Thomas Toys—USA
Timpo—GB
Tiny Car—GB
Tomica—J
Tomica-Aviva—J
Tomite
Tomte—N
Tomy—J
Tonka—USA
Tootsietoy—USA
Topping Models—USA
Toy & Tackle—USA
Trafficast—GB
Triton-Lego—NL
Triang—GB
Triang-Minic—GB

Triang Spot On—IRL
Trol—BR
Tru-Scale (Carter)—USA
Trumm, Eldon—USA
Trumm-DeBaillie—USA
Tudor Rose—GB

U S U D—CS
Umex—A
Underwood Engineering
    (Fun-Ho)—NZ
Universal-Coreg—HK
Uralec

VM Miniatures—GB
Valley Patterns—USA
Varney Copy Cat—GB
Vendeuvre—F
Verve—I
Vilmer—DK
Vindex—USA
Vinyl Line—D
Viproduct Victory—GB

W & K—GB
W & T Rowland—GB
Wader—D
Wannatoys—USA
Wardie BJW—GB
Warner, Gerry—USA
Werner Wood &
    Plastic Co—USA
The Wheel Works—USA
White's Model Shop—USA
Wiking—D
Willis Finecast—GB
Winross—USA
Winter, Don—USA
Wooden Toys—USA
Woodlands Scenics—USA
Wyandotte—USA

Y-(Diapet)—J
Yaxon—I
Yaxon S.A.—ZA
Yoder, Weldon—USA
Yonezawa (Diapet)—J

Z T S Plastik—POL
Ziss R.W. Modelle (D)
Zylmex—HK

Numerical Codes

Numerical Code	Tractor Manufacturer	Letter Code	Numerical Code	Tractor Manufacturer	Letter Code	Numerical Code	Tractor Manufacturer	Letter Code
010300-000	Allis Chalmers	AC	060113-000	Famulus	FAM	140900-000	New Idea	NI
010718-000	Agrale	AGR	060118-000	Farmtoy	FAR	141500-000	Norm	NO
010800-000	Arthur Hammer	AH	060514-000	Fendt	FEN	141518-000	Normag Zorge	NOR
011212-000	Allchin	ALL	060518-000	Ferguson	FER	142106-000	Nuffield	NUF
011600-000	Agripower	AP	060901-000	Fiat	FIA	151209-000	Oliver	OLI
011818-000	Arrow	ARR	061301-000	Field Marshall	FMA	151616-000	Opperman	OPP
012119-000	Austin	AUS	061504-000	Foden	FOD	152015-000	Odero Terni Orlando	OTO
012200-000	Averling & Porter	A&P	061718-000	Ford	FOR	152301-000	Owatonna	OWA
012205-000	Avery	AV	061719-000	Fordson	FOT	160114-000	Panhard	PAN
012220-000	Avto	AVT	061723-000	Fowler	FOW	160518-000	Perplex	PER
020111-000	Baker	BAK	061809-000	Frick	FRI	160719-000	P.G.S.	PGS
020120-000	Bates	BAT	062214-000	Funkstorgrad N	FUN	161518-000	Porsche	POR
020121-000	Bautz	BAU	070118-000	Garnett	GAR	180514-000	Renault	REN
020401-000	Beauce-Flandre	BEA	070508-000	Gehl	GEH	181519-000	Rosengart	ROS
020412-000	Belarus	BEL	071802-000	Graham Bradley	GRB	182113-000	Rumley	RUM
020907-000	Big Bud	BIG	072109-000	Guidart	GUI	190113-000	Same	SAM
021200-000	Blackhawk	BL	072112-000	Guldner	GUL	190114-000	Satoh	SAT
021201-000	Blaw Knox	BLA	080114-000	Hanomag	HAN	190301-000	Scale Models	SCA
021322-000	B M Volvo	BMV	080115-000	Hanomag-Barreiros	HAO	190522-000	SFV	SFV
021512-000	Bolinder Munktell	BOL	080519-000	Hesston	HES	190805-000	Sheppard Diesel	SHE
021600-000	Buffalo-Pitts	BP	080914-000	Hindustan	HIN	190912-000	Silver King	SIL
022118-000	Burrell	BUR	081512-000	Holder	HOL	190913-000	Simplicity	SIM
022211-000	Bukh	BUK	081514-000	Honda	HON	191513-000	Someca	SOM
030118-000	Carraro	CAR	082102-000	Huber	HUB	192005-000	Steiger	STE
030119-000	Case	CASE	082519-000	Hyster	HYS	192018-000	Steyr	STY
030120-000	Cassani	CAT	090803-000	IH	IHC	192301-000	Swaraj	SWA
030801-000	Chaseside	CHA	091407-000	Ingeco	ING	200112-000	Talyor	TAL
031201-000	Clark-Bobcat	CLA	091420-000	International	INT	200518-000	Terratrak	TER
031204-000	Cletrac	CLE	091909-000	Iseki	ISK	200913-000	Timberjack	TIM
031503-000	Cockshutt	COC	100302-000	JCB	JCB	200920-000	Titan	TIT
031512-000	Colorado	COL	100400-000	John Deere	JD	201512-000	Toldi	TOL
031514-000	Continental	CON	110918-000	Kirovec	KIR	201725-000	Toyota	TOY
031515-000	Co-op	COO	111801-000	Kramer	KRA	201801-000	Track Marshall	TRA
031521-000	County	COU	112102-000	Kubota	KUB	202107-000	Tugster	TUG
031522-000	Coventry-Climax	COV	120113-000	Lamborghini	LAM	202309-000	Twin City	TWI
032102-000	Cub Cadet Corp.	CUB	120114-000	Landini	LAN	211819-000	Ursus	URS
040106-000	DAF	DAF	120126-000	Lanz	LAZ	211820-000	Urtrack	URT
040109-000	Dain	DAI	120525-000	Leyland	LEY	220112-000	Valmet	VAL
040200-000	David Brown	DB	120919-000	Lister	LIS	220514-000	Vendeuvre	VEN
040521-000	Deutz	DEU	130106-000	Massey-Ferguson	MAF	220518-000	Versatile	VER
040920-000	Ditch Witch	DIT	130108-000	Massey-Harris	MAH	220903-000	Vickers	VIC
041505-000	Doe Triple-D	DOE	130114-000	Man	MAN	220905-000	Vierzon	VIE
041714-000	Dong Fong	DON	130115-000	Manitou	MAT	230112-000	Wallis	WAL
042120-000	Dutra	DUT	130502-000	Mercedes-Benz	MEB	230501-000	Weatherhill	WEA
050107-000	Eagle	EAG	130509-000	Meili	MEI	230519-000	Westrak	WES
050118-000	Ebro	EBR	130518-000	Merlin	MER	230809-000	White or WFE	WHI
050120-000	Eaton Yale	EAT	130903-000	Michigan	MIC	230812-000	Wheel Horse	WHL
050315-000	Economy	ECO	130920-000	Mitsubishi	MIT	230820-000	Whitlock	WHT
050903-000	Eicher	EIC	131300-000	Minneapolis-Moline	MM	231512-000	Wolseley	WOL
051819-000	Ertl	ERT	131514-000	Monarch	MON	250112-000	Yale-Eaton	YAL
051903-000	Escort	ESC	131913-000	M. S. & M. Co.	MSM	250114-000	Yanmar	YAN
060108-000	Fahr	FAH	132108-000	Muir-Hill	MUH	260520-000	Zetor	ZET
060109-000	Fairbanks-Moorse	FAI	140800-000	New Holland	NH	260525-000	Zettlemeyer	ZEY

061719-024, -025, -026 (F) 3¾"—4¾" A.C. WILLIAMS

090803-126A FARMALL F-12 1/16 A T & T
COLLECTABLES

061719-056 (F) 5½" A. C. WILLIAMS

100400-150 GP 2-ROW 1/16 A T & T COLLECTABLES

011212-002 7-32 1/76 ABS—MODELS

090803-126B FARMALL F-14 1/16 A T &
COLLECTABLES

160114-001 ?? 1/43 AR

030119-066 RC 1/16 A T & T COLLECTABLES

Individual Tractors

090803-163 F-30 1/16 A T & T COLLECTABLES

100400-151A B 1/16 A T & T COLLECTABLES

100400-081 (AR) 1/16 A T & T COLLECTABLES

190805-002 SD-3 1/16 A T & T COLLECTABLES

031503-001 30 1/16 ADVANCED PRODUCTS

031515-001 (E-3) 1/16 ADVANCED PRODUCTS

031503-002 (540) 1/16 ADVANCED PRODUCTS

060518-005 (20) 1/12 ADVANCED PRODUCTS

031515-002 (E-3) 1/16 ADVANCED PRODUCTS

040200-006 VAK CROPMASTER 1/72 AIRFIX

030118-002 6500 1/20 AGFA DI FAVERO

020121-002 300 1/30 ALLEMAGNE PLASTY (Cursor)

060518-008 (TE-30) 1/20 AIRFIX

050903-006 1955 1/20 PENDING

060518-009 30 1/25 ALLEMAGNE—PLASTY

090803-069 560 1/16 ALPS TOYS

192005-013 SERIES II 1/16 GARY ANDERSON

010300-009 (C) 1/12 AMERICAN PRECISION

151209-027 88 ROW CROP 1/16 GARY ANDERSON

010300-083 440 1/16 GARY ANDERSON

010300-003 (U) 3'' ARCADE

010300-004 (U) 5'' ARCADE

061718-002 (9N) 6-1/2'' ARCADE

010300-005 (WC) 6'' ARCADE

061718-003 (9N ROW CROP) 3-1/4'' ARCADE

010300-006 (WC) 7'' ARCADE
061718-001 (9N) 6-1/2'' ARCADE

061719-001 (F) 3-7/8''   061719-002 (F) 4-3/4'' ARCADE
061719-003 (F) 5-3/4'' ARCADE

 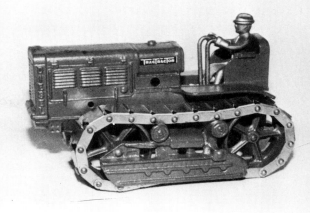

061719-004 (F) 6''   061719-005 (F) 6'' ARCADE   090803-003 TRAC-TRACTOR (TD-40) 1/16 ARCADE

090803-001A (10-20) 1/16 090803-001B (10-20) 1/16 ARCADE   090803-003 TRAC-TRACTOR (TD-40) 1/16 ARCADE

090803-002A FARMALL REGULAR 1/16 ARCADE   090803-004 TRAC-TRACTOR (TD-18) 1/16 ARCADE

090803-002B FARMALL REGULAR 1/16 ARCADE   090803-005 M 4-1/4'' ARCADE

090803-006 M 5-1/4" ARCADE

100400-003 (A) 1/16 ARCADE

090803-007 M 1/16 ARCADE

151209-001 (70) 1/25 ARCADE

090803-008 A 1/12 ARCADE

151209-002 70 1/16 ARCADE

160518-001 7300 1/16 ARNOLD

220112-001 138-4 1/50 ARPRA

120126-016 BULLDOG 5" APTHKYA

010300-008 (WC) 1/16 AUBURN RUBBER

010718-001 4300 1/25 ARPRA LTD

071802-001 32 HP 4" AUBURN RUBBER

090803-022 (M) 4'' AUBURN RUBBER

131300-003B Z 1/32 AUBURN RUBBER

100400-010 (MT) 1/20 AUBURN RUBBER

131300-004A (R) 1/16 AUBURN RUBBER
131300-004B (R) 1/16 AUBURN RUBBER

131300-003A Z 1/32 AUBURN RUBBER

151209-012 70 1/20 AUBURN RUBBER

030801-001 ?? 1/87 AUGUPLAS

090803-122 400 1/16 ROY LEE BAKER

080115-001 R 440 1/87 AUGUPLAS-MINICARS

090803-141 1933 F-12 1/16 ROY LEE BAKER

100400- 1/64 DENNIS BAKER

130108-030 20K 1/28 BP

090803-167 M-TA 1/5 ROY LEE BAKER

020111-001A 21-75 SPECIAL 1/25 IRVIN

090803-168 F-12 1/5 ROY LEE BAKER

060518-001A (30) 1/43 BENBROS

100400-102A 7020 1/32 ROY LEE BAKER
100400-102B 7020 1/32 ROY LEE BAKER

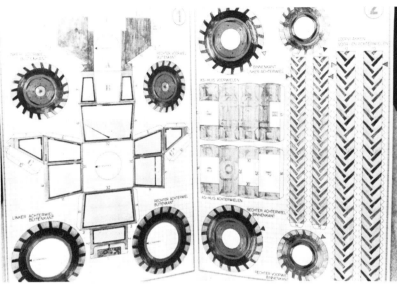

060514-015 FAVORIT TURBOMATIK BIJAGE VAN HET

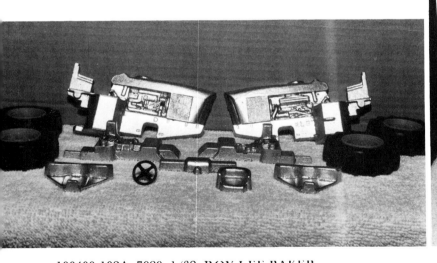

100400-102A 7020 1/32 ROY LEE BAKER

100400-155 (D) 5-1/2'' BANTHRICO

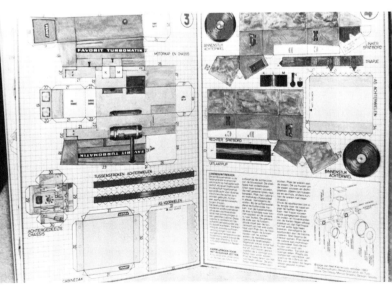

061719-055 (F) / HALF TRACKS 8'' BING

061719-054 (F) 8'' BING

040521-013B DX-110 1/32 BRITAINS LTD

061719-063 MAJOR   BLUE BOX

040521-013D DEUTZ—FAHR DX-110 1/32 BRITAINS LTD

130115-001 FORK LIFT 1/25 BOURBON

021322-005B BM 2654 1/32 BRITAINS LTD

021322-005A BM2654 1/32 BRITAINS LTD.
The mower is made by Scaledown, not Britains Ltd.

040521-013E DEUTZ-FAHR DX-92 1/32 BRITAINS LTD.

040521-013E DEUTZ-FAHR DX-92 1/32 BRITAINS LTD.

061718-038A 5000 1/32 BRITAINS LTD.

040521-013F DX-110 INDUSTRIAL 1/32 BRITAINS LTD

061718-038B 5000 1/32 BRITAINS LTD.

060901-010D 880/HALF TRACKS 1/32 BRITAINS LTD

SAME AS BOTTOM LEFT.

060901-010E 880 DT 1/32 060901-010A 880DT 1/32 BRITAINS LTD

061718-038C 5000 1/32 BRITAINS LTD.

061718-039 5000 1/43 BRITAINS LTD.

061718-067B 7710 1/32 BRITAINS LTD

061718-040B 6600/ROPS & SCRAPER 1/32
BRITAINS LTD.

061718-068 (6600) 1/32 BRITAINS LTD

061718-040A 6600 1/32 BRITAINS LTD.

061718-067A TW-20 1/32 BRITAINS LTD

061718-067C 7710 INDUSTRIAL 1/32 BRITAINS LTD

061719-028 (E27N) 1/76 BRITAINS LTD.

061719-029A (E27N) 1/32 061719-029B (E27N) 1/32
BRITAINS LTD.

100302-010 520.4 LOADALL 1/32 BRITAINS LTD.

061719-030A POWER MAJOR 1/32 BRITAINS LTD.

061719-032A 5000 SUPER MAJOR 1/32 BRITAINS LTD.
061719-033 5000 1/32 BRITAINS LTD.

061719-030B POWER MAJOR 1/32 BRITAINS LTD.
061719-031B SUPER MAJOR 1/32 BRITAINS LTD.
061719-031A SUPER MAJOR 1/32 BRITAINS LTD.

130106-022B 135 INDUSTRIAL 1/32 BRITAINS LTD.

061719-031C INDUSTRIAL MAJOR 1/32 BRITAINS LTD.

130106-022C 135 1/32 BRITAINS LTD.

130502-002A UNIMOG 1/32 BRITAINS LTD.

130106-022A 135 1/32 BRITAINS LTD.
130106-023A 595 1/32 BRITAINS LTD.

130502-002C UNIMOG 1/32 BRITAINS LTD.

130106-023C 595 1/32 BRITAINS LTD.

130106-061A 2680 1/32 BRITAINS LTD.

130502-002B UNIMOG 1/32 BRITAINS LTD.

130502-006A 1500 MB 1/32 BRITAINS LTD.

130502-006B 1500 MB 1/32 BRITAINS LTD.

061719-060A E27N 1/32 BROWN'S MODELS
061719-060B E27N 1/32 BROWN'S MODELS

180514-023 TX 145-14 1/32 BRITAINS LTD

061719-067A (F) 1/32 BROWN'S MODELS

061719-051 N STANDARD 1/32 BROWN'S MODELS

060518-016A   FERGUSON—BLACK   1/32
060518-016B   FERGUSON—BLACK   1/32   BROWN'S MODELS

061719-060A E27N 1/32 BROWN'S MODELS

100400-071A, B, C   B 1/32 BROWN'S MODELS

100400-071D, E, F B 1/32 BROWN'S MODELS

090803-098 844-S 1/30 BRUDER

060514-014 F255GT 9" BRUDER

040521-021C 06 SERIES 8" BRUDER

040521-017 06 SERIES 1/87 BRUDER

040521-021A 06 SERIES 8" BRUDER

060514-009C TURBO. FARMER 308LS 1/16 BRUDER
040521-021B 06 SERIES 8" BRUDER

130502-005 MB 1/20 BRUDER

060901-014B (25R) 1/42 BUDGIE

130502-005 1300 MB 1/20 BRUDER

130106-073 (175) BULLI

130502-009 MB TRAC 1/87 BRUDER

060901-001 FIAT CONCORD 700-S 1/43 BUBY

120113-002A ?? 1/16 CAVALLINO

120113-002B ?? 1/16 CAVALLINO

180514-004A R 30-40 1/16 CIJ

061719-036 DEXTA 1/16 CHAD VALLEY

180514-001 R 30-40 1/32   180514-002 E-30 1/32 CIJ

060518-003A (30) 1/16 CHAD VALLEY

180514-004B R 30-40 1/16 CIJ

180514-022 E-30 1/55 CIJ

061719-034 (E27N) 1/16 CHAD VALLEY
061719-035 (DDN) MAJOR 1/16 CHAD VALLEY

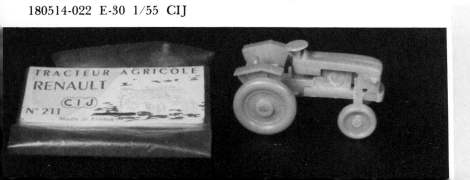

061719-079 E27N 1/43 CHAD VALLEY

030119-026 12-20 CROSSMOUNT 1/16 KEN CONKLIN

022211-001 D-30 DIESEL 1/43 CHICO TOYS

030119-029A 580D CK 1/35 CONRAD

130106-011 (35) 1/32 CHICO TOYS

190113-003 LEONE 70 1/12 CO-MA

030119-029B 580D CK SILVER ANN. 1/35 CONRAD

050903-003B 3105A 1/35 CONRAD

030119-028 580C CK 1/35 CONRAD

010300-040 FORKLIFT 1/36 CORGI

030119-055 4890 1/35 CONRAD

030119-021A CASE—DB 1412 1/32 CORGI

060518-015 TE-20 1/35 CONRAD

130106-014B & C 165 1/43 CORGI

030119-021B CASE—D.B. 1412 1/32 CORGI
061719-043A POWER MAJOR 1/43 CORGI
061719-043B POWER MAJOR 1/43 CORGI

061718-041A, B, C 5000 SUPER MAJOR 1/43 CORGI

130106-014D 165 1/43 CORGI

061719-044A & B POWER MAJOR 1/43 CORGI

130106-015 165 1/78 CORGI (JR)

130106-012A & B 50-B 1/43 CORGI

130106-013A, B, C 65 1/43 CORGI

021322-002 800 1/66 CORGI JR

260520-001 5511 1/80 CORGI (JR)

100400-006A WATERLOO BOY (N) 1/16 CHARLES COX

061718-021 4040 1/12 CRAGSTAN

100400-005 FROELICH 1/16 CHARLES COX

061718-020 4040 1/12 CRAGSTAN

061718-019 1841 INDUSTRIAL 1/12 CRAGSTAN

061718-022 4040 HD
1/12 CRAGSTAN

061719-048 (DEXTA) 1/25 CRESCENT

060518-020 (30) ? CRESCENT

040521-025 D-40L 1/21 ? CURSOR

060514-006 FAVORIT 1/43 CURSOR

050903-008 ?? 20 cm CURSOR

060514-002 F 250 GT 1/25 CURSOR

060108-001 ?? 1/25 CURSOR

060108-004 ?? 1/50 CURSOR

060514-003 F 250 GT 1/43 CURSOR   060514-004 ?? 1/43

060514-002 & -003 F 250GT 1/25 & 1/43 CURSOR

060514-007 FARMER 2 1/20 CURSOR

072112-002 TOLEDO 1/20 CURSOR

060514-011B DIESELROSS 6 PS 1/32 CURSOR

080114-002 PERFECT 400 1/32 CURSOR

060514-023 FAVORIT 15 cm CURSOR

081512-003 C-500 1/27.5 CURSOR

072112-001 SPRINTER I 1/32 CURSOR

080114-002 PERFECT 400 1/32 CURSOR

080114-002 PERFECT 400 1/32 CURSOR

081512-001 CULTITRAC A55 1/30 CURSOR

080114-008A B16 1/50 CURSOR

081512-002A CULTITRAC A60 1/27.5 CURSOR

080114-009 R16 1/43 ??

080114-010 R12 1/50 ??

081512-002B CULTITRAC A60 1/27.5 CURSOR

111801-001 ?? 12 CM CURSOR

180514-009 ?? 1/87 CURSOR

180514-015 86 1/12 CURSOR?

130502-025 UNIMOG 425 1/87 CURSOR

180514-016 SUPER 5 1/87 CURSOR

141518-001 ZORGE 1/70 CURSOR

161518-007 DIESEL STANDARD 5'' CURSOR

180514-019 MASTER 1/25 CURSOR

060514-020 FARMER 10'' DBGM

010300-007 (WC) 1/16 DENT

112102-004 L2402DT SUNSHINE 1/23 DIAPET

061719-014 (F) 5-3/4'' DENT

112102-006 M79500T 1/20 DIAPET

021322-004 T 650 1/50 Y—DIAPET

250114-001 YM 3110 1/25 DIAPET

060901-032 (880) 1/16 DICKIE (SPIEIZEUG)

030119-047 D-STANDARD 1/16 LYLE DINGMAN

030119-046A (DC) 1/16 LYLE DINGMAN

030119-067 LA 1/16 LYLE DINGMAN

030119-047 D—STANDARD 1/16 LYLE DINGMAN

030119-046B (DC) LPG 1/16 LYLE DINGMAN

100400-101 (H) 1/16 LYLE DINGMAN

151209-026A 770 1/16   151209-026B 880 1/16 LYLE DINGMAN

090803-127B B 1/16 LYLE DINGMAN
090803-127A A 1/16 LYLE DINGMAN

021201-006 CRAWLER 1/43 DINKY

021201-001 CRAWLER 1/43 DINKY

021201-002 CRAWLER 1/43 DINKY

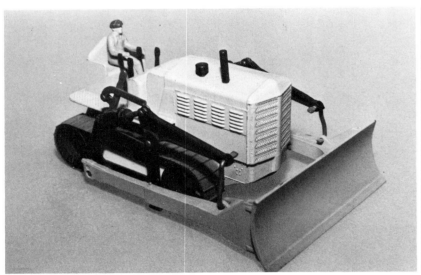

021201-001B CRAWLER 1/43 DINKY

030119-010 CASE—D.B. 995 1/43 DINKY

061301-001A & B SERIES I 1/43 DINKY

040200-002 990 1/43 DINKY
040200-003 990 SELECTOMATIC 1/43 DINKY

061719-015A (F) 1/43 DINKY

130108-002 (55) 1/80 DINKY-DUBLO

120525-001A   120525-001B 384 1/43 DINKY

042120-004 D4K 6'' DUBENA

130108-001B   130108-001A (44) 1/43 DINKY

132108-001B 2WL 1/43 DINKY

060901-002A 550 1/35 DUGU

060901-004A 550 060901-004B 600 1/36 ZISS R W MODELLE

190113-002 CENTAURO DT 1/15 DUGU

182113-002 OIL PULL (20-30) 1/16 ALVIN EBERSOL

130106-053 (165) 8-1/2'' DURAVIT

090803-074A 844 1/43 ELIGOR

090803-128A 8-16 1/16 ALVIN EBERSOL

140800-001 1/2 HP 1/16 ALVIN EBERSOL
090803-128B 8-16 1/16 ALVIN EBERSOL

130108-045 (44) 4-1/2'' EMPIRE MADE

130108-046 (44) 2'' EMPIRE MADE

010300-030 (D—SERIES) 12" EMPIRE

130108-046 (44) 2" EMPIRE MADE

John Deere

HERE IS A REAL SALES BUILDER CR-1070-E 10-3

ALLIS-CHALMERS WD-45 Kelton Corp. TL1771.

010300-015 (WC) 6'' ERTL

010300-019B 190 1/16 ERTL

010300-016A (D-SER. I) 1/16 ERTL

010300-018 LGT 1/16 ERTL

010300-016B (D-SER. II) 1/16 ERTL

010300-017B B-112 1/16 ERTL   010300-017A B-110 1/16 ERTL

010300-016D (D-SER. IV) 1/16 ERTL

010300-020 190 XT 1/16 ERTL

010300-022A BIG ACE 1/16 ERTL

010300-023B 200 1/16 ERTL

010300-021 190 XT SERIES III LANDHANDLER 1/16
ERTL

010300-024A 7030 1/16 ERTL

010300-023A 200 1/16 ERTL

010300-024B 7040 1/16 ERTL

010300-024C 7050 1/16 ERTL

010300-025B FIAT—ALLIS 12-GB 1/25 ERTL

010300-024D 7060 1/16 ERTL

010300-032A 7045 1/16 ERTL

010300-025A 12-G 1/25 ERTL

010300-032B 7060 1/16 ERTL

010300-032C 7080 1/16 ERTL

010300-041 8070 1/64 ERTL

010300-032D 7080 1/16 ERTL

010300-043A 8030 1/16 ERTL

010300-039 8550 1/32 ERTL

010300-043B 8010 1/16 ERTL

010300-044A 4 W-305 1/32 ERTL
010300-044B 4 W-305 1/32 ERTL

030119-007B 1070 AK 1/16 ERTL

010300-043C 8030 1/16 ERTL

030119-005B 1030 1/16 ERTL

030119-005A 930 CK 1/16 ERTL

030119-006 1030 CK 1/16 ERTL

030119-007C 1070 AK DEMONSTRATOR 1/16 ERTL

030119-008C 1370 AK 504 TURBO 1/16 ERTL

030119-007D 1070 AK 451 DEMONSTRATOR 1/16 ERTL

030119-008B 1270 AK 451 1/16 ERTL

030119-008A 1270 AK 451 1/16 ERTL

030119-008C 1370 AK 504 TURBO 1/16 ERTL

030119-008D (1370) AGRI—KING 1/16 ERTL

030119-023 (AGRI—KING) 1/64 ERTL

030119-008E (1370) AGRI—KING 1/16 ERTL

030119-025A 2390 1/16 ERTL

030119-008F SPIRIT OF 76 1/16 ERTL

030119-008G SPIRIT OF 76 1/16 ERTL

030119-025B 2590 1/16 ERTL

030119-025C 2390 COLL. SER. 1/16 ERTL

030119-034A   030119-034B 4890 1/32 ERTL

030119-025E 2390 030119-025F 2590 RECALL 1/16 ERTL

030119-038A 2290 1/32 ERTL

030119-025D 2590 COLL. SER. 1/16 ERTL

030119-038B 2290 COLL. SER. 1/32   030119-053 2294 1/32 ERTL

030119-025G 2590 1/16 ERTL

030119-039A 1690 1/32 ERTL

030119-041 2590 1/64 ERTL

030119-051A 2594 COLL. SER. 1/16 ERTL

030119-038C 2294 1/32 ERTL

030119-039B 1690 COLL. SER.   030119-040C 1690 1/32 ERTL

030119-052A 3294 COLL. SER. 1/16 ERTL
030119-060B 4894 1/32 ERTL

031503-005A 1850 1/16 ERTL
031503-005B 1850 1/16 ERTL

040200-007 1690 1/32 ERTL

061718-026A 4000 1/12 ERTL

040920-001A 4010 1/50 ERTL

061718-025 4000 1/12 ERTL

040920-001B 4010 1/50 ERTL

061719-018 (F) 1/16 ERTL

061718-026B 4400 INDUSTRIAL 1/12 ERTL

061718-027A 4600 1/12 ERTL

061718-028A 5550 INDUSTRIAL 1/12 ERTL

061718-029C 8600 1/12 ERTL

061718-029A 8000 1/12 ERTL

061718-029D 9600 1/12 ERTL

061718-029B 8000 1/12 ERTL

061718-030 145 HYDRO LGT 1/12 ERTL

061718-050 7700 1/12 ERTL

061718-072A   061718-072B FW-60 1/64 ERTL

061718-051 9700 1/12 ERTL

061718-074 555 INDUSTRIAL 1/32 ERTL

061718-052B TW-20 1/64 ERTL

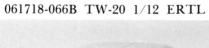

061718-066B TW-20 1/12 ERTL

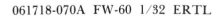

061718-069 TW-20 1/32 ERTL

061718-070A FW-60 1/32 ERTL

061718-086A LGT 12—LIM.ED. 1/12 ERTL

090803-027B CUB CADET 122 1/16 ERTL

061718-075A 7710 1/16 ERTL

090803-026 (M) 1/16 ERTL

061718-077 TW-15 1/32 ERTL

090803-025 (M) 1/16 ERTL

090803-028A CUB CADET 125   090803-029B CUB CADET 126 1/16 ERTL

090803-030B CUB
CADET 129 1/16
ERTL

090803-030D CUB CADET 129 1/16 ERTL

090803-034A 400 1/16 ERTL

090803-031A CUB CADET 1650 1/16 ERTL

090803-034A 400 1/16 ERTL

090803-032A 240 1/16 ERTL

090803-034B 400 1/16 ERTL

090803-033 340 INDUSTRIAL 1/16 ERTL

090803-035 404 1/16 ERTL

090803-036A & B 404 1/16 ERTL

090803-039A & B 544 1/16 ERTL

090803-036B 404 1/16 ERTL

090803-038E & F 560 1/16 ERTL

090803-038A 460 & -038B 560 1/16 ERTL

090803-038F 560 1/16 ERTL

090803-039F (2644) INDUSTRIAL 1/16 ERTL

090803-040B FLYING FARMALL 1/16 ERTL

090803-041A FARMALL 656 1/32 ERTL

090803-041B INTERNATIONAL 656 1/32 ERTL

090803-044C 1256 TURBO 1/16 ERTL

090803-038H 560 TOY FARMER 1/16 ERTL

090803-044A 856 1/16 ERTL

090803-043A 806 1/16 ERTL

090803-043B 806 1/16 ERTL

090803-044B 1026 HYDRO 1/16 ERTL

090803-042A (666) 1/32 ERTL

090803-042B (666) 1/32 ERTL

090803-045C 1066 TURBO 1/16 ERTL

090803-044D 1456 TURBO 1/16 ERTL

090803-045D 1066 TURBO 1/16 ERTL

090803-045A 966 HYDRO 1/16 ERTL

090803-046 1206 TURBO 1/16 ERTL

090803-045E 1466 1/16 ERTL

090803-049B 1086 1/16 ERTL

090803-048B 1466 1/64 ERTL

090803-079 1086 1/64 ERTL

090803-049C 1586 1/16 ERTL

090803-049A 886 1/16 ERTL

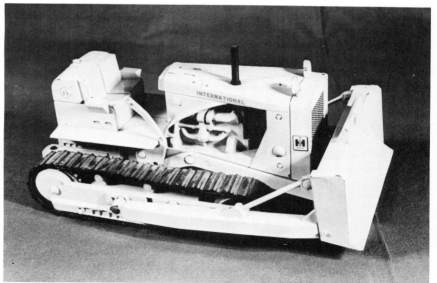

090803-050 TD-25 1/25 ERTL

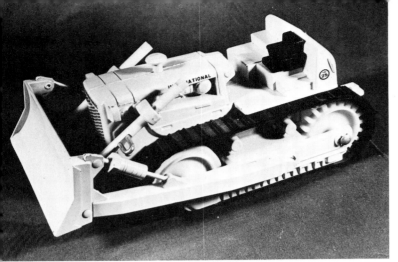

090803-051 TD-25 1/25 ERTL

090803-082D 6388 2 + 2 1/16 ERTL

090803-053A 3414 1/16 ERTL

090803-082A 3588 (2 + 2) 1/16 ERTL

090803-053B 3444 1/16 ERTL

090803-054 (3400) 1/16 ERTL

090803-091A 090803-091E CUB CADET (682) 1/16 ERTL

090803-100A 5288 1/16 ERTL

090803-100A 5288 1/16 ERTL

090803-100B 5088 1/16 ERTL

090803-088 McCORMICK FARMALL 1/64 ERTL

090803-086 TITAN 10-20 1/64 ERTL

090803-087 McCORMICK 10-20 1/64 ERTL

090803-085B IHC TYPE A 1/64 ERTL

090803-089 FARMALL (H) 1/64 ERTL

090803-100D 5288 1/16 ERTL

090803-091B CUB CADET 682 1/16 ERTL
090803-091D CUB CADET (682) 1/16 ERTL

090803-109A 5088 1/64 ERTL

090803-100D

090803-102A 270 BACKHOE-LOADER 1/64 ERTL

090803-105A 560 PAYLOADER 1/80 ERTL

090803-106A 640 EXCAVATOR 1/64 ERTL

090803-103A 350 PAYHAULER 1/80 ERTL

090803-100E 5288 1/16 ERTL

090803-101 412B 1/64 ERTL

090803-100G 5488 1/16 ERTL

090803-109B 5088 1/64 ERTL

090803-109A

090803-136 (3088) 1/16 ERTL

090803-109A

090803-157A 300 1/16 ERTL

090803-125A 2 + 2 3588 1/64 ERTL

090803-125B 2 + 2 6388 1/64 ERTL

090803-110 1086 1/16 ERTL

090803-136 (3088) 1/16 ERTL

090803-139 6388 2 + 2 1/40 ERTL

090803-169A 7488 2 + 2 1/16 ERTL

100400-021B (A) 1/16 ERTL

090803-172 BACKHOE 1/16 ERTL

100400-025 110 1/16 ERTL

100400-020 (D) 1/16 ERTL

100400-017A (730), 018A (4010), 019 (4430), 013 WATER-
LOO BOY, 014 (17), 106 (60) 1/64 ERTL

100400-023A (B) 1/16 ERTL

100400-022A (A) 1/16 ERTL

100400-024 (60) 1/16 ERTL

100400-028A (430) 1/16 ERTL

100400-026A 140 1/16 ERTL

100400-029A (620) 1/16 ERTL
100400-029B (620) 1/16 ERTL

100400-026B, 026C, 026D, 026E 140 1/16 ERTL

100400-027A (400) 1/16 ERTL

100400-032A (2040) 1/16 ERTL

100400-030A (630) 1/16 ERTL

100400-029A (620) 1/16 ERTL

100400-031A (2030) 1/16 ERTL
100400-031B (2030) 1/16 ERTL

The first major casting variation representing the gasoline engine tractor was produced from 1961 to sometime in 1964. It had a generator, three-point-hitch, upside down U-shaped hand grip, three levers on the top left side of dash, one lever on top right side of dash. All four wheels were die-cast metal with narrow 3/4'' rear rims and narrow 5/16'' front rims. This one was assigned the number 100400-033.

The second major casting variation appeared in 1964 and ran until 1969. Representing a diesel engine with a generator, it had two 3/16 inch long fuel filters on the left side of the engine, three levers on the top left side of dash and one lever on the top right side of dash. In addition there was a lever on each cowl side, below the dash. There were eight minor variations numbered 100400-034C thru -034J.

The third major casting variation represented the diesel engine tractor with alternator instead of generator. First made in 1969, it completed this series in 1972. It had two large 5/16 inch fuel filters on the left side. There was a lever on both the top left and top right side of the dash. There were no levers on either side of the cowl. Four minor variations numbered 100400-035A thru -035Q rounded out this one.

Side view of JOHN DEERE 3010-4010 & 3020-4020 variations.

100400-033A: 3010-4010 with small front (2) and rear tires (5), metal front and rear wheels, three-point-hitch (4), large P.T.O. lever, no fuel filters (3), one transmission filter on the left side (6), generator (9), fenders are screwed on. All 3010-4010 tractors have rubber front and rear tires. Note: Fenders could be removed so that corn picker could be mounted on tractor.

100400-033B: Same but without three-point-hitch.

Left: Plastic wheels with small rubber front tires. Right: Metal wheels with small front tires.

Left: No fuel filters. Center: Inverted U-shaped grab bar. Right: Large PTO lever.

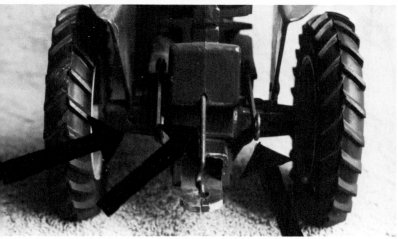

Left: *Hole for mounting corn picker, fender screwed on.* Center: *Upper or center three-point-hitch arm.* Right: *Lower or side three-point-hitch arm.*

100400-034C: 3020-4020 with small front and rear tires, metal front and rear wheels, three-point-hitch, three hydraulic and one shift lever, small levers on both sides (8), two transmission filters (7), generator, and fenders screwed on. All 3020-4020 tractors have rubber rear tires.

Left: *Small rear wheel and tire.* Right: *Large rear wheel and tire.*

100400-034D: Same but without three-point-hitch, bottom three-point-hitch holes are open but no hole in tractor for upper center three-point-hitch arm (10).

100400-034E: Small front and rear tires, rear metal wheels, plastic front wheels, three-point-hitch, three hydraulic and one shift lever, small lever on each side of cowl, small fuel filters on left side (8), two transmission filters, generator and fenders screwed on.

100400-034F: Same but without three-point-hitch, bottom three-point-hitch holes are open but no hole for center three-point-hitch arm (10).

Left: *Rear tire size differences.* Right: *Single transmission filter on bottom of tractor.*

Left: *Two transmission filters on bottom of tractor.* Right: *Rear tire size differences.*

100400-034G: Small tires on front and rear, plastic front tires, plastic rear wheels (10), three-point-hitch, three hydraulic and one shift levers, small lever on each side of cowl, small fuel filters on left side, two transmission filters, generator and fenders screwed on.

Left: *Two small oil filters on left side of engine.* Right: *Small lever on side of cowl and small grab bar.*

100400-034H: Same but without three-point-hitch, bottom three-point-hitch holes open but no hole for center three-point-hitch arm. Note—No more three-point-hitches on any later tractors.

100400-034I: Small rear tires (5), extra large wide rubber front tires, plastic front wheels (2), small metal rear wheels, no three-point-hitch (10), three hydraulic and one shift levers (15), small lever on each side of cowl, small fuel filters on left side, two transmission filters, generator (9) and fenders screwed on.

Left: *Small lever on side of cowl and large grab bar on right side.* Right: *Generator on right side of engine.*

100400-034J: Small front and rear tires (6), plastic front tires, plastic front and rear wheels, no three-point-hitch, two hydraulic and one shift levers, small lever on each side of cowl (8 & 9), small fuel filters on left side, two transmission filters (7), generator and fenders screwed on. Note—All tractors with small fuel filters have closed axle housing on bottom. All tractors with large fuel filters have open axle housing on bottom, axle being visible from one mounting to the other when tractor is inverted.

Left: No side three-point-hitch arms. Center: No center three-point-hitch arm or hole for it. Right: Rear plastic wheel. Notice ribs.

100400-035A: Small front and rear tires, plastic front tires, plastic wheels front and rear, no three-point-hitch, one hydraulic and one shift levers (16), large fuel filters on left side (12), no side levers, two transmission filters, alternator (11), fenders riveted on. All three-point-hitch holes closed.

100400-035B: Same except right side three-point-hitch hole open and center hole closed.

100400-035C: Same except both side three-point-hitch holes open and center hole closed.

Lower left: No small lever on side of cowl. Right: Alternator on right side of engine.

100400-035D: Small front and rear tires, plastic front tires, plastic front and rear wheels, no hydraulic lever, no shift lever, large fuel filters on left side (12), no side levers, two transmission filters and alternator (11). All three-point-hitch holes are closed. Upper left: Large grab bar on right side.

100400-035E: Small tires on front, plastic front tires and plastic front wheels.

100400-035F: Same except front tires are extra large and wide rubber (13).

100400-035G: Same except rear wheels are plastic and front tires are large plastic.

100400-035H: Large plastic front tires, plastic front wheels, large wide rear tires (7), plastic rear wheels, one hydraulic and one shift lever, no three-point-hitch, large fuel filters on left side (12), no side levers (12), two transmission filters, alternator, fenders riveted on and all three-point-hitch holes are closed.

Right: Large grab bar and no small side lever on left side of cowl.

100400-035I: Same except one three-point-hitch hole open and center hole and other side closed. Left: Large oil filters on left side of engine.

100400-035J: Same, both lower three-point-hitch holes open and center hole closed.

100400-035K: Large plastic front tires with plastic rims, large wide rubber rear tires (6), plastic rear rims, wide front end (13), no three-point-hitch, three hydraulic and one shift levers, small lever on each side of cowl, small fuel filters on left side, two transmission filters (7), generator, fenders screwed on and all three-point-hitch holes closed.

Left: Wide front axle with wide tires. Right: Narrow or row crop front end with wide tires.

100400-035N: Same except both side three-point-hitch holes open and center hole closed.

100400-035O: Large plastic front tires with plastic rims, wide front end (13), large wide rubber rear tires with plastic rims, no three-point-hitch, no hydraulic levers, one shift lever, large fuel filters on left side, two transmission filters, large muffler, alternator, fenders and ROPS riveted on, left three-point-hitch hole is open, right hole and center hole closed, SMV sign on rear (14).

100400-035P: Same except one shift lever and all three-point-hitch holes are closed.

100400-035Q: Same except one shift lever, left three-point-hitch hole open and right and center holes closed.

100400-035L: Large plastic front tires with plastic rims, wide front ends (13), large wide rear tires with plastic rims, no three-point-hitch, three hydraulic levers and one shift lever, small lever on each side of cowl, small fuel filters on left side (8), two transmission filters, generator (9), fenders and ROPS riveted on, left three-point-hitch hole open, right and center holes closed.

Lower left: SMV (Slow Moving Vehicle) sign or emblem. Center: "Bolt on" ROPS (Roll Over Protection System). Right: ROPS.

100400-035M: Large plastic front tires with plastic rims, wide front end, large wide rubber rear tires with plastic rims (7), no three-point-hitch, one hydraulic and one shift lever (16), large fuel filters on left side, two transmission filters, fenders riveted on, left three-point-hitch hole open, right and center holes closed.

Three hydraulic levers on left side of dash and one shift lever on right side.

Single hydraulic lever on left side of dash and one shift lever on right side of dash.

119

100400-036A (4430) 1/16 ERTL

100400-037A (4430) 1/32 ERTL

100400-036B (4430) 1/16 ERTL

100400-037C (4440) 1/32 ERTL

100400-036D (4440) 1/16 ERTL

100400-036G 4250 TOY FARMER 1/16 ERTL

1004000-037C Left, low hitch, no cab. Right, high hitch, cab.

100400-039B 5020 1/16 ERTL

100400-037D 50 SERIES 1/32 ERTL

100400-039A No front axle brace.

100400-038 4430 1/25 ERTL

100400-039C 5020 1/16 ERTL Angle braces.

100400-039A 5020 1/16 ERTL

100400-039D 5020 1/16 ERTL Solid angle braces.

100400-042A (40) CRAWLER 1/16 ERTL

100400-043 (440) INDUSTRIAL CRAWLER 1/16 ERTL

100400-040A (7520) 1/16 ERTL

100400-041A (8630) 1/16 ERTL

100400-039E 5020 1/16 ERTL

100400-042C (420) CRAWLER 1/16 ERTL

100400-042B (40) INDUSTRIAL CRAWLER 1/16 ERTL

100400-049A 1/25 ERTL, Also available with ROPS.

100400-048 (JD-310) 1/25 ERTL

100400-065 (4430) 1/64 ERTL

100400-044A (440) INDUSTRIAL TRACTOR 1/16 ERTL

100400-050 (570) GRADER 1/25 ERTL

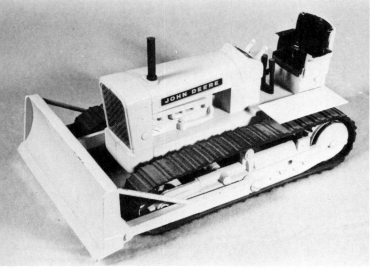

100400-045 (1010) INDUSTRIAL CRAWLER 1/16 ERTL

100400-052B (869) SCRAPER 1/25 ERTL

100400-047 (JD-350) 1/16 ERTL

100400-076 (4440) 1/64 ERTL

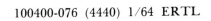

100400-053 LOG SKIDDER 1/16 ERTL

100400-073 3140 1/32 ERTL

100400-061A SKID STEER LOADER 1/16 ERTL

100400-080 TRAILFIRE 440 1/12 ERTL

100400-072B 3140 1/32 ERTL

100400-085A (950) COMPACT UTILITY 1/16 ERTL

100400-085A Without wedge to hold muffler in place (early variation) decal backwards, "STRIPES" should be to the front of the tractor. -085A Wedge to hold muffler in place.

100400-087A 8650 COL. SER. 1/16 ERTL

100400-088A 4850 COL. SER. 1/16 ERTL

100400-098 BACKHOE 1/32 ERTL

130106-029A 175 DIESEL 1/16 ERTL

100400-099 BACKHOE 1/64 ERTL

130106-029A 175 DIESEL 1/16 ERTL

100400-108 50 SERIES ROW CROP 1/64 ERTL

100400-031A (2030) 1/16 ERTL

130106-029B 3165 1/16 ERTL

130106-031C 1150 1/16 ERTL

130106-030 275 DIESEL 1/16 ERTL

130106-031D 1150 V-8 1/16 ERTL

130106-032A 1105 1/16 ERTL

130106-034A 1155 1/16 ERTL

130106-032B 1105 1/16 ERTL

130106-034B 1155 1/16 ERTL

130106-033A 1155 1/64 ERTL

130106-034C 1155 1/16 ERTL

130106-035 1155 1/25 ERTL

130106-057 670 1/20 ERTL

130106-035 1155 1/25 ERTL

130106-036 1155 1/25 ERTL

130106-037A 590   130106-037B 595 1/16 ERTL

130106-051 4880 1/64 ERTL

130106-058A 270 1/16 ERTL

130106-059A 698 1/20 ERTL

130106-042A 2775 1/20 ERTL

130106-042B 2805 1/20 ERTL

130106-058B 270 COLL. SER. 1/16 ERTL

130106-059B 698 1/20 ERTL

130106-052B 4880 1/32 ERTL

131300-014A G-1000 VISTA 1/16 ERTL

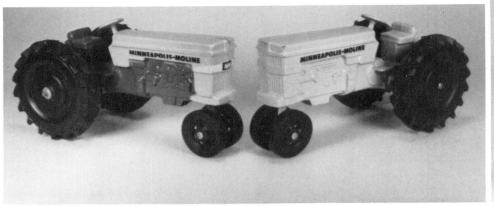

131300-011A   131300-011B (602) 1/25 ERTL

131300-014B (G-1000 VISTA) 1/16 ERTL

131300-012C THERMOGAS (602) 1/25 ERTL

131300-016 G-1355 1/16 ERTL

131300-015A MIGHTY MINNIE 1/16 ERTL

131300-012A (602 LPG)   131300-013 (ROW CROP LPG) 1/25 ERTL

151209-017C 1800 (SERIES C)
151209-017D 1800 FWA 1/16 ERTL
151209-017K 1855 1/16 ERTL

151209-017A   1800 (SERIES A), -017B   1800
(SERIES B) 1/16 ERTL

151209-017L 1855 1/16 ERTL

152301-002A MUSTANG 1/16 ERTL

151209-017J 151209-017M WHITE 1855 1/16 ERTL
151209-017H WHITE 1855 1/16 ERTL

152301-001 MUSTANG 1/16 ERTL
192005-003A PANTHER 1/32 ERTL

192005-003B COUGAR III ST 251
192005-003D PANTHER ST 310 1/32 ERTL

192005-007A COUGAR 1/64 ERTL

192005-003C PANTHER ST 310 1/32 ERTL

192005-003E INDUSTRIAL 1/32 ERTL

192005-007B INDUSTRIAL 1/64 ERTL

192005-003F INDUSTRIAL 1/32 ERTL

061718-087 (5000) 1/48 ESCI
061718-087 The tractor (061718-087) is a part of this plastic model kit.

042120-002 D4K 1/87 ESPEWE PLASTICART

060901-005A 780   060901-005B 880DT 1/43 FORMA—PLAST

211820-001C KT 50 1/87 ESPEWE

120113-001A R 1056 1/43 FORMA-PLAST

060113-001 ?? 1/87 ESPEWE

061719-023 (F) 9-1/4'' EZRA BROOKS

060901-011A 780 1/13 FORMA—PLAST

060901-011B 880 DT 1/13 FORMA—PLAST

090803-118 300   090803-119 350 1/16 PETE FREIHEIT

1004000-093 (G) 1/16 PETE FREIHEIT

090803-116B SUPER H 1/16 PETE FREIHEIT

100400-093 (G) 1/16 PETE FREIHEIT

090803-115 SUPER M-TA 1/16 PETE FREIHEIT

090803-120 230 1/16 PETE FREIHEIT

030119-043 VAC 1/16 PETE FREIHEIT

090803-117 140 1/16 PETE FREIHEIT

060518-010 (30) 4-1/2" FUN—HO

130108-006 (44) 3" FUN-HO

060518-011 (35) 6-1/2" FUN—HO

130108-007 (44) 6-1/2" FUN-HO
090803-064 (M) FUN-HO 11-1/2" FUN-HO

061719-019 (F) 5-1/2" FUN-HO

061719-020 SUPER MAJOR 1/87 FUN-HO

151209-014 70 ORCHARD 1/25 FUN—HO

061718-035 (4000) 2-1/4"   061718-036 (4000) 3-3/4" FUN—HO

090803-140 D-430 1/32 G & S ERZEUGNIS   040521-009F INTRAC 2005 1/28 GAMA

060901-015 880 1/19 G.P.   040521-009B INTRAC 2005 1/28 GAMA

061718-049 (4000) 1/43 GALANITE   040521-008A D 100 06A 1/29 GAMA

040521-008A D 100 06A 1/29 GAMA   040521-009A INTRAC 2005 1/28 GAMA

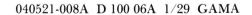

040521-012 GAMA—DEUTZ 1/19 GAMA

061718-031 (4000) 7'' GAMA

040521-020 D10006 GAMA

061718-034A (4000) 1/43 GAMA

040521-023 DX 230 POWERMATIC 1/12 GAMA

060514-010 FAVORIT 620LS TURBO. 1/32 GAMA

060901-003D 25R 1/43 MINI—GAMA

130106-055 2640 1/32 GAMA

161518-001 ?? 1/43 GAMA

031201-002 533 HYDROSTATIC 1/24 GAMA

130106-010 35 1/43 GAMDA

061718-045 (145) LGT GAY TOYS

130502-004 MB TRAC 1300 1/20 GAMA

060901-003B 25R 1/43 MINI—GAMA
060901-003D 25R 1/43 MINI-GAMA

061718-044 (4000) 10'' GAY TOYS

130502-003 MB TRAC 1300   GESCHA

120114-005 L35 HOT BULB 1/16 GIANCO

030119-019A 380B CK 1/16 GESCHA

030118-001E BULL 34 (88.4) 1/24 SCAME—GIODI
030118-001F MASTER 33 (88.4) 1/24 SCAME—GIODI

031201-001A   031201-001B M-700 1/24 GESCHA

030118-001B 88.4 1/24 SCAME—GIODI

130108-021 (44) 1/87 GITANES

010300-002 (A) 1/12 ROBERT GRAY

010300-037 (A) 1/12 ROBERT GRAY

090803-075A FARMALL F-30 8'' ROBERT GRAY

090803-010B (10-20) 1/16 ROBERT GRAY

090803-084A IHC GAS ENGINE 1/16 ROBERT GRAY

090803-010A (10-20) 1/16 ROBERT GRAY

021600-001 ?? 1/25 ROBERT GRAY

090803-084B IHC GAS ENGINE 1/16 ROBERT GRAY

090803-076A McCORMICK W-9 8-1/2'' ROBERT GRAY

100400-066B GAS ENGINE 1/16 ROBERT GRAY

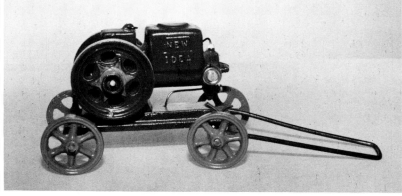

140900-001B GAS ENGINE 1/16 ROBERT GRAY

100400-009A (A-GP) 1/12 ROBERT GRAY

151209-015A HART-PARR (28-44) 1/12 ROBERT GRAY

130108-022A CHALLANGER 1/12 ROBERT GRAY

140900-001A GAS ENGINE 1/16 ROBERT GRAY

151209-015B HART-PARR (28-44) 1/12 ROBERT GRAY

151209-016A 70 RC 1/16 ROBERT GRAY

151209-016B 151209-016C 70 STANDARD 1/16 ROBERT GRAY

010300-052 RC, -053 W 1/16 GUBBELS

202309-001 (60-90) 1/16 ROBERT GRAY

010300-056 WC STYLED, -057 WC UNSTYLED 1/10
GUBBELS

060901-027 (550) 120113-003 (653) 1/50 GRISONI GIOCATTOLI

010300-054 WF UNSTYLED, -055 WF STYLED 1/1
GUBBELS

061719-077 (DEXTA) 1/45 GRISONI GIOCATTOLI

010300-049 A, -050 UC, -051 U 1/12 GUBBELS

010300-058 B 1/10, -059 C 1/12, -060 CA 1/12 GUBBELS

010300-067 D-19 PROPANE, -068 D-19 DIESEL, -069 D-19 GAS 1/12 GUBBELS

010300-061 D-10 1/12 GUBBELS

010300-070 D-21, -071 D-21 SERIES II 1/12 GUBBELS

010300-062 D-12 1/12 GUBBELS

010300-063 WD 1/12, -064 WD-45 LPG 1/12, -065 WD-45 DIESEL 1/10, -066 WD-45 GAS 1/12 GUBBELS

010300-073 220, -072 210 1/12 GUBBELS

131300-037, 090803-164 M 1/12 GUBBELS

061718-024 (4000) 1/56 GUISVAL

040109-001 ALL WHEEL DRIVE 1/16 FRANK HANSEN

120525-002 (384) 1/12 H P PLAST (NYRHINEN)

040109-001 ALL WHEEL DRIVE 1/16 FRANK HANSEN

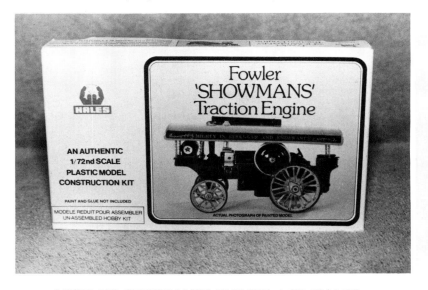

061723-002 SHOWMAN'S ENGINE 1/72 HALES

061718-089 (8000) 1/64 HALLMARK

040521-014B D6206 1/32 HAUSSER

040521-019 F1L 514 1/20 HAUSSER

040521-015 (D-15) 1/32 HAUSSER ELASTOLIN

130114-001 ?? 1/32 HERBART

040521-016 INTRAC 2003 1/32 HAUSSER

161518-010 STANDARD 1/32 HERBART

090803-093 423 1/28 HAUSSER

130106-060 2680 1/24 HELLER BOBCAT (HUMBROL)

130114-001 ?? 1/32 HERBART

012205-005 (A) 1/16 TOM HOFFMAN

131300-024 GTB 1/16 WALLY HOOKER

061718-013 (8N) 1/24 HUBLEY

130108-044 44 1/16 JIM HOSCH

061718-016A (4000) 1/12 HUBLEY

061718-058 5000 SUPER MAJOR   HOVER

061718-078 5000 8'' HOYUK

061719-006 (F) 3-3/4''   061719-007 (F) 3-1/2'' HUBLEY

061718-014 (960) H 1/10 HUBLEY

061718-015A 961 POWERMASTER 1/12 HUBLEY

061718-016B 4000 1/12 HUBLEY

061718-015B 961 POWERMASTER 1/12 HUBLEY

061718-017B 6000 1/12 HUBLEY

061718-015C SELECT—O—SPEED 1/12 HUBLEY

061718-017A 6000 1/12 HUBLEY

147

061718-018B   COMMANDER 6000   1/12   HUBLEY—
GABRIEL

061719-008 (F) LOADER 8-1/2'' HUBLEY

061718-018A COMMANDER 6000 1/12 HUBLEY

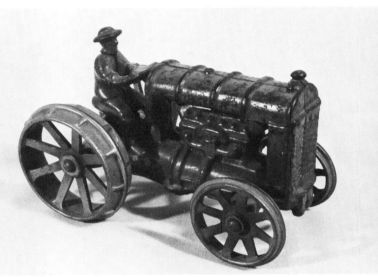

061719-009 (F) 5-1/2'' HUBLEY

061719-008 (F)/LOADER 8-1/2'' HUBLEY

082102-001 STEAM ROLLER 1/25 HUBLEY

090803-021 (M) 1/12 HUBLEY

260520-005C CRYSTAL 12045 1/43 IGRA

151209-003 70 ORCHARD 1/25 HUBLEY

260520-005B CRYSTAL 12045 1/43 IGRA

082102-004 STEAM ROLLER 3-1/4" HUBLEY

260520-005A CRYSTAL 12045 1/43 IGRA

131514-001 Small Hubley, -002 Medium Hubley, -003 Large
Hubley, -004 5-1/2" OTT & PTW

010300-038 7045 1/32 IMAI

020111-001A 21-75 SPECIAL 1/25 IRVIN

061718-065 9700 PULLER 1/32 IMAI

030119-009B STEAM ROLLER 1/16 IRVIN

012200-004 UNDERMOUNT 1/25 IRVIN
012200-003 AVERY 4½" IRVIN
030119-009A ?? 1/25 WHITE; PETERSON; IRVIN

050107-001   GASOLINE ENGINE, -001   GASOLINE
ENGINE, 060109-003 Z-ENGINE 1/10 IRVIN

061809-001 16 HP 1/25 WHITE: PETERSON: IRVIN

220905-001 401 1/30 JRD

082102-002A RETURN FLU, -002B STEAM ROLLER
1/25 BRUBAKER; WHITE; PETERSON; IRVIN

120126-007 BULLDOG 1/43 JAYA

182113-003 OIL PULL 1/25 IRVIN

060901-028 BD 20 1/20 ITES

010300-036A (WC) 1/16 EARL JERGENSEN

010300-036B (WC) 1/16 EARL JERGENSEN

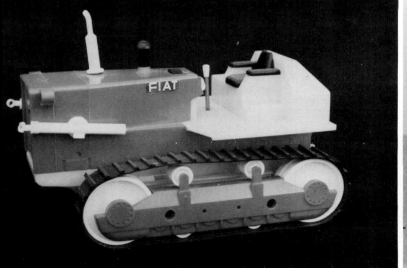

090803-113 FARMALL (C) 1/16 JOUEF

100400-092 (MC) 1/16 K & G SAND CASTING

130106-017A 65X 1/43 JUE

130106-019 3366 1/43 JUE

100400-091A (M) 1/16 K & G SAND CASTING

260520-009 CRYSTAL 8011 1/18 KDN

260520-008A 25 1/20 KDN

061809-001 16 HP 1/25 WHITE; PETERSON; IRVIN

220905-001 401 1/30 JRD

082102-002A RETURN FLU, -002B STEAM ROLLER
1/25 BRUBAKER; WHITE; PETERSON; IRVIN

120126-007 BULLDOG 1/43 JAYA

182113-003 OIL PULL 1/25 IRVIN

010300-036A (WC) 1/16 EARL JERGENSEN

060901-028 BD 20 1/20 ITES

010300-036B (WC) 1/16 EARL JERGENSEN

010300-036B (WC) 1/16 EARL JERGENSEN

050118-002 470 1/43 JOAL

030119-022 (CC) 1/16 EARL JERGENSEN

130106-025A   130106-025B 165 1/43 JOAL

100400-167 (D) 1/16 EARL JERGENSEN

130106-050 EBRO 470 1/43 JOAL

050118-001 6100 1/43 JOAL

030119-004A CASOMATIC 1/16 JOHAN

030119-004B 800 CASOMATIC 1/16 JOHAN

061719-022A (F) FARM 1/87 JORDAN PRODUCTS

090803-112A FARMALL (C) 1/30 JOUEF

090803-112B FARMALL (C) 1/30 JOUEF

090803-113 FARMALL (C) 1/16 JOUEF

090803-111 FARMALL (C) 1/65 JOUEF

090803-113 FARMALL (C) 1/16 JOUEF

090803-113 FARMALL (C) 1/16 JOUEF

100400-092 (MC) 1/16 K & G SAND CASTING

130106-017A 65X 1/43 JUE

130106-019 3366 1/43 JUE

100400-091A (M) 1/16 K & G SAND CASTING

260520-009 CRYSTAL 8011 1/18 KDN

260520-008A 25 1/20 KDN

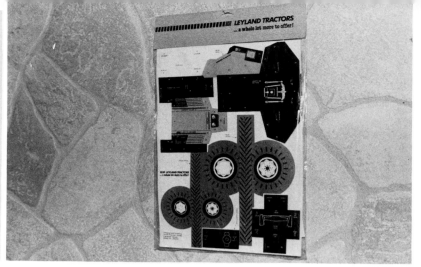

260520-008B 25 1/20 KDN

120525-003 804 1/18 KARRAN PRODUCTS LTD.

100400-004 (D) 5'' KANSAS TOY

011818-001 AGRICULTURAL 1/12 KARSLAKE

060514-021A FARMER 1/12 KARKURA

061723-004 CLASS 27 PLOW. ENG. 1/72 KEIL KRAFT

130108-039 44 1/20 KARKURO

031503-006 30 1/16 KEMP PRODUCTS LTD.

090803-138 (M) 1/16 KEMP PRODUCTS LTD.

090803-009 (10-20) 1/20 KILGORE

061719-012 (F) 6-1/2'' KENTON

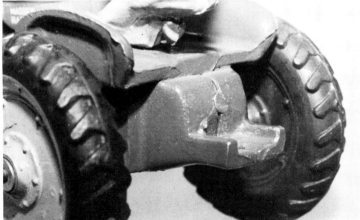

061719-082 (F) 5-1/4'' KENTON

061719-010 (F) 4'' KILGORE
061719-011 (F) 5-1/8'' KILGORE

130108-018 44 1/16 THE KING COMPANY

156

130108-018 DETAIL, THE KING COMPANY

010300-077 18-30 1/10 KRUSE

130108-019 44 1/16 THE KING COMPANY

130108-019 DETAIL, THE KING COMPANY

010300-075 WD 1/10 KRUSE

030119-064 C 1/10 KRUSE

010300-076 6-12 1/10 KRUSE

010300-078 20-35 1/10 KRUSE

030119-065 L 1/10 KRUSE

010300-079 A 1/10 KRUSE

090803-165 F-30 1/10 KRUSE

010300-080 WC 1/10 KRUSE

030119-057 CC 1/10 MARVIN KRUSE

090803-166 F-20 1/10 KRUSE

100400-116 B 1/10 KRUSE

100400-112 D 1/10 KRUSE

100400-117 G 1/10 KRUSE

100400-113 GP STANDARD TREAD 1/10 KRUSE

100400-118 "A W" STYLED 1/10 KRUSE

100400-114 GP WIDE TREAD 1/10 KRUSE

100400-119 "A" STYLED 1/10 KRUSE

100400-115 A 1/10 KRUSE

100400-120 "B" STYLED 1/10 KRUSE

100400-121 "G" STYLED 1/10 KRUSE

100400-125 JOHN DEERE-LINDEMANN 1/10 KRUSE

100400-122 "D" STYLED 1/10 KRUSE

100400-128 80 1/10 KRUSE

100400-123A "L" STYLED 1/10 KRUSE

100400-124 "L" UNSTYLED 1/10 KRUSE

100400-127 60 1/10 KRUSE

100400-126 50 1/10 KRUSE

100400-129 R 1/10 KRUSE

100400-133 820 1/10 KRUSE

100400-130 520 1/10 KRUSE

100400-134 430 1/10 KRUSE

100400-131 620 1/10 KRUSE

100400-135 530 1/10 KRUSE

100400-132 720 GAS 1/10 KRUSE

100400-136 630 1/10 KRUSE

100400-137 730 GAS 1/10 KRUSE

100400-148 GAS ENGINE 1/10 KRUSE

182113-005 RUMLEY 6 1/10 KRUSE

100400-138 830 1/10 KRUSE

112102-003 ?? 1/17 KUBOTA-JIGYOSHA

192018-002 180a or 80 or 280 1/30 KUNSTOFF HERBERT FITZEK

192018-002 180a or 80 or 280 1/30 KUNSTOFF HERBERT FITZEK

090803-015A SUPER C 1/16 LAKONE

110918-001 ?? 1/43 LENINGRAD

090803-015B SUPER C 1/16 LAKONE-CLASSIC

021201-003 CRAWLER
1/43 MOKO—LESNEY

060518-002 (30) 3-1/2'' MOKO—LESNEY

090803-016B 230 1/16 LAKONE

100400-056A (3020) 1/32 LEE TOYS
100400-056B (3020) 1/32 LEE TOYS

010300-027 SCRAPER PAN 7-1/2'' LIONEL

130108-009 745D 1/15 MOKO-LESNEY

011212-001 7-32 1/80 LESNEY

061718-023A   061718-023B (4000) 1/56 LESNEY

061719-041 POWER MAJOR 1/62 LESNEY

061723-001 BIG LION 1/80 LESNEY

061719-042 SUPER MAJOR 1/42 LESNEY

090803-066A   090803-066B B-250 1/37 LESNEY

030119-016 (1000D) 1/82 LESNEY

030119-017 (1000D) 3-1/2'' LESNEY

100400-057A   100400-057B JD-LANZ 700 1/61 LESNEY

151209-013 77 STANDARD 1/16 LINCOLN SPECIALITIES

130106-024 165 1/43 LESNEY

130108-016C 44 1/16 LINCOLN SPECIALITIES

132108-002A 161 1/50 LESNEY

130108-016B 44 1/16 LINCOLN SPECIALITIES

230501-001 (12H) 1/57 LESNEY

130108-016B 44 1/16 LINCOLN SPECIALTIES

130108-015B 44 STANDARD 1/20 LINCOLN SPECIAL-
ITIES

031503-004 30 1/16 LINCOLN SPECIALITIES

130108-015A LINCOLN SPECIALITIES

130108-015B 44 STANDARD 1/20 LINCOLN SPECIAL-
TIES

100400-011 (A) 1/16 LINCOLN SPECIALITIES

090803-024 (M) 6-1/2'' LINCOLN SPECIALITIES

130108-016A 44 1/16 LINCOLN SPECIALITIES

130108-023B (745) 1/20 LINCOLN MICRO MODELS

130108-023A (745) 1/20 LINCOLN MICRO MODELS

010300-026A HD-15 4'' LIONEL

130108-012 (44) 1/20 MAJOR-MODELS

010300-027 SCRAPER PAN 7-1/2'' LIONEL

061719-084A (N) 1/87 LINES BROS.
061719-084B COVENTRY CLIMAX 1/87 LINES BROS.

130106-007 35 1/32 LION—MOLBERG

040200-008 (1690) 1/32 LONE STAR

030119-054 (1594) 1/32 LONE STAR

061718-091 7610 1/32 LONE STAR

130106-070 675 1/32 LONESTAR

061718-012A (NAA ROADMASTER) 1/32 DCMT
LONESTAR

100400-156 (A) GP 1/16 LUCHT-FRIESEN

130106-062 (2680) 1/32 LONESTAR

061719-049 (DEXTA) 5-1/4" LUCKY

090803-097 946 1/32 LONE STAR

061718-008 (8N) 1/12 M.P.C.

230812-001 LGT 1/24 MPC

050903-004 KOIGSTIGER 1/15 MICHAEL SEIDEL

050903-001A KONIGSTIGER 1/18 MS TOY

061719-039 MAJOR 1/20 M.W.

161518-011 ALLGAIER POR.A-133 1/32 MS

152015-001 C 25 R 1/20 MACHPI-BOLOGNA

152015-001 C25R 1/20 MACHPI BOLOGNA

180514-010 3000 1/32 MAJORETTE

152015-002 C 25 C 1/20 MACHPI-BOLOGNA

030119-061 CC 1/12 MARBIL

260520-007 SUPER 1/25 MACHPI CASCHI

061718-042A 5000 1/55 MAJORETTE

010300-042A U 1/16 MARBIL

130108-029C 744 1/16 MARBIL

010300-042B U 1/16 MARBIL

080114-007 ?? 5-3/4" MARUSAN TOYS

031521-001 FORDSON E27N CRAWLER 1/16 MARBIL

080114-015A SS 100 1/43 MARKLIN

061719-071 E27N 1/16 MARBIL

120126-002 BULLDOG 1/43 MARKLIN

120126-002, -003 BULLDOG 1/43 MARKLEN

090803-077 FARMALL 6-1/2" MARX

090803-019 (M) DIESEL FIXALL 1/12 MARX

051903-004 335 1/43 MATTELL—MINI

090803-063 (M) 4" MARX

090803-063 (M) 4" MARX

061718-082 3600 1/43 MATTELL MINI

050903-002B GOODEARTH 1/20 MAXWELL

051903-002   051903-003 335 1/20 MAXWELL

090803-092 B-275 MAHINDRA 1/20 MAXWELL

061718-057 5000 SUPER MAJOR 1/43 MAXWELL

192301-001A (735) 1/20 MAXWELL

061718-083 3600 1/20 MAXWELL

260520-003 HMT 2511 1/25 MAXWELL

080914-001 ?? 1/20 MAXWELL

130502-007 UNIMOG 29 CM MEHANO TEKNICA

160719-002 ROMA 38 18'' MEHANO TEHNIKA

080114-011 (ST 100) 1/40 MERCURY

120126-008 BULLDOG 1/200 MERCATOR

120114-001A R-400, -001B R-500 1/43 MERCURY

010300-028 CRAWLER 1/80 MERCURY

120114-001A R-400, -001B R-500 1/43 MERCURY

060901-012A (780) 1/87 MERCURY

130106-009 35 1/43 MERCURY

040200-005 CROPMASTER 1-1/2'' MERIT

130108-011 (44) 1/43 MICRO-MODELS

060518-007 (TE-20) 5-1/2'' METTOY—CASTOYS

060518-014 20 DEMONSTRATOR 1/10 MILLS BROS.

061719-050 (DEXTA) 5-1/4'' MIC

060901-007A FL-10 1/24 MINI—AUTO

060518-012 (35) 1/32 MICRO—MODELS

061719-046 SUPER MAJOR 1/43 MINALUXE

142106-003 (UNIVERSAL) 1/16 MINIC

082519-003 FORKLIFT TEKNO

130106-018A 275 1/43 MINIMAC

080519-003A 980DT 1/64 MINI-TOYS

080519-003B 980 1/64 MINI-TOYS

080519-003C 1180DT 1/64 MINI-TOYS

080519-003D 1180 1/64 MINI-TOYS

080519-003E 1180DT 1/64 MINI-TOYS

080519-003F 1180 1/64 MINI-TOYS

080519-003J 130-90 DT 1/64 MINI-TOYS

080519-003G 1180 TURBO 1/64 MINI-TOYS

080519-003M 130-90 1/64 MINI-TOYS

080519-003H 100-90 DT 1/64 MINI-TOYS

080519-003K 130-90 1/64 MINI-TOYS

080519-003I 100-90 1/64 MINI-TOYS

080519-003L 130-90 DT 1/64 MINI-TOYS

130108-027 (44) 4" MITOPLAST

090803-013 TRAC-TRACTOR 1/24 MOD-AC MFG.

180514-007 651 or 652 1/13 MONT BLANC

030119-003A (SC) 1/16 MONARCH PLASTIC
030119-003B (SC) 1/16 MONARCH PLASTIC
090803-095 624 DIESEL 11" MONT BLANC

180514-021A 651 1/13 MONT BLANC

060901-021 702 (50TH ANNIVERSARY) 1/12 MOPLAS

051903-001 335 1/25 MORGAN MILTON LTD
051903-002 335 1/20 MAXWELL

100400-104A   100400-104B 3185 1/43 NPS

061718-053 3600 1/20 MORGAN MILTON LTD.

130106-028 450-S 1/50 NZG

130106-008A 35 1/25 MORGAN MILTON LTD.

090803-143B 3180 1/43 NPS

030119-012 (1412) 1/25 NZG

030119-011 CASE—D.B. 1412 1/25 NZG

030119-013 (2670) TK 1/40 NZG

030119-024 850—B CRAWLER 1/35 NZG

061718-047 4550 INDUSTRIAL 1/35 NZG

031514-001 CD8 1/43 NOREV

130106-027 300 1/50 NZG

070508-001 4610 1/25 NZG

180514-006A R-86 1/43 NOREV

031514-001 CD8 1/43 NOREV

061719-013 (F) 3-1/2" NORTH & JUDD

180514-006B R-86 1/43 NOREV

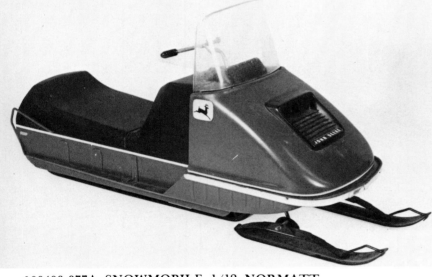

100400-077A SNOWMOBILE 1/12 NORMATT

061718-092 (JUBILEE NAA) 1/32 NOSTALIG

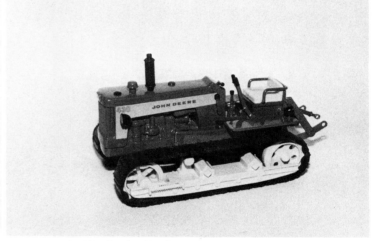

100400-159 430 CRAWLER 1/16 GEORGE NYGREN

100400-078 JDX SNOWMOBILE 1/12 NORMATT

060901-025 FORKLIFT 1/50 OLD CARS

060901-023 FIAT—ALLIS FR-20 1/50 OLD CARS

090803-011A (10-20) 1/16 OLD TIME TOYS

060901-008 640 1/36 OLD CARS

060901-002C 640 1/35 DUGU

130108-043 55 1/16 OLD TIME COLLECTABLES

100400-009C (A-GP) 1/12 OLD TIME COLLECTABLES

130106-069 55 1/16 OLD TIME COLLECTABLES

100400-007 (D) 1/16 OTT; PTW

061719-017 (F) 6'' OTT, PTW

182113-001A OIL PULL (20-35) 1/16 OLD TIME TOYS

010300-046 BIG ORANGE 750 ML PACESETTER

130108-014 745 1/38 P. M. I.

010300-047 BIG ORANGE 50 ML PACESETTER

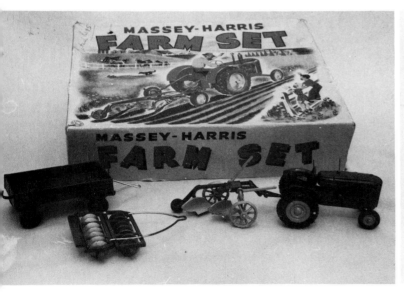

130108-014 745 1/38 P.M.I.

030119-062 CASE (2594) 750 ML PACESETTER

030119-063 CASE (2594) 50 ML PACESETTER

090803-146 2 + 2 200 ML PACESETTER

090803-133 (5288) 50 ML PACESETTER

090803-145 2 + 2 1 LITER PACESETTER

090803-134 (5288) 750 ML PACESETTER

100400-089A (4440) 1 LITER PACESETTER

184

100400-090 (4440) 200 ML PACESETTER

192005-010 PANTHER CP-1400 200 ML PACESETTER

100400-164 PACESETTER (8650) 200 ML PACESETTER

061301-004 SERIES II 1/32 BRIAN PARKS

192005-009 PANTHER CP-1400 1.0 LITER PACESETTER

010300-035 WC 1/16 DENNIS PARKER

030119-042 400 1/16 DENNIS PARKER

131300-018 R 1/16 DENNIS PARKER

090803-170 SUPER M-TA 1/16 DENNIS PARKER

100400-109 4020 1/16 DENNIS PARKER

100400-064 A-GP 1/16 DENNIS PARKER

100400-105 (D) 1/16 DENNIS PARKER

131300-022 UB 1/16 DENNIS PARKER

082102-002A RETURN FLU 1/25 BRUBAKER; WHITE; PETERSON; IRVIN

060113-002 ?? 1/16 PIKO

090803-011B (10-20) 1/16 PIONEER TRACTOR WORKS

030119-002B (L) 1/16 PIONEER TRACTOR WORKS

061719-068 (F) 1/16 PIONEER TRACTOR WORKS

030119-059 SC 1/16 BURKHOLDER—PTW

060109-001 Z-ENGINE 1/16 ARCADE

060109-002 Z—ENGINE 1/16 OTT; PTW

100400-002 GAS ENGINE 1/16 VINDEX

100400-008 GAS ENGINE 1/16 OTT; PTW
090803-012B FARMALL REGULAR 1/16 PIONEER
TRACTOR WORKS

100400-007 (D) 1/16 OTT; PTW

130108-024 (55) 6-1/2'' PLASTICUM

130106-056 (135) 1/64 PLAYART DICKIE

112102-002A L-1500 1/23 DIAPET

112102-005 KUBOTA (L2402DT) 1/23 PLAYART DICKIE

090803-153 SUPER C 1/16 PIONEER TRACTOR
WORKS

131514-004 ?? 5-1/2'' OTT & PTW
182113-001B OIL PULL (20-35) 1/16 PIONEER
TRACTOR WORKS

131300-019 25HP POWER UNIT 1/16
PLOW BOY TOYS

188

200518-001 ?? 6-1/2" PLOW BOY TOYS

061719-045 MAJOR 1/41 POLITOYS

230519-001 ?? 7" PLOW BOY TOYS

100400-157 JD-850 1/43 PRECISION CRAFT PEWTER

190113-001 LEONE 70 1/65 POLISTIL

061718-064 8700 1/41 POLISTIL

080114-012A R45 1/87 PREISER

061718-046 8600 8-1/2" PROCESSED PLASTIC
PRODUCTS

010300-010 (WD-45) 1/16 PRODUCT MINIATURE

061718-005B 600 1/16 PRODUCT MINIATURE

010300-012 HD-5 1/16 PRODUCT MINIATURE

090803-014A (M) 1/16 PRODUCT MINIATURE

061718-004 8N 1/12 PRODUCT MINIATURE
061718-005A NAA JUBILEE 1/12 PRODUCT MINIATURE

061718-006 900 1/12 PRODUCT MINIATURE
090803-014C (M)/STARS 1/16 PRODUCT MINIATURE

090803-017 UD-24 POWER UNIT 1/16 PRODUCT
MINIATURE

160719-001 420 ROMA 1/9 PROTAR

090803-018A TD-24 CRAWLER 1/16 PRODUCT
MINIATURE

160719-001 420 ROMA 1/9 PROTAR

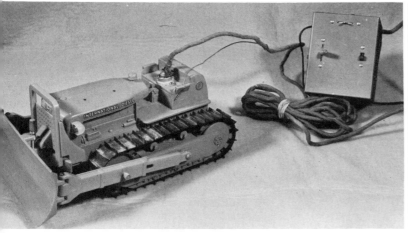

090803-018B TD-24 CRAWLER 1/16 PRODUCT
MINIATURE

200913-001 240E 1/12 PROMOTION G. BELANGER

061718-009A (8N) 5'' QUIRALU

031512-001 ?? CRAWLER 5'' QUIRALU

142106-002 (UNIVERSAL) 1/15 RAPHAEL LIPKIN

130106-021 175 DIESEL 1/20 REINDEER

130108-010A (745) 1/20 RAPHAEL LIPKIN

060518-019 TE-20 1/43 REPLICA DIE CAST

130108-010B 745 1/20 RAPHAEL LIPKIN

130106-020A 135 DIESEL 1/38 REINDEER

100400-059B 100400-059A JD-LANZ (300/500) 5'' REX

100400-060 JD-LANZ 1010 5-1/2'' REX

120126-004 BULLDOG 1/21 REX

130108-020B 44 1/16 REUHL PRODUCTS

151209-019 (70) 1/16 E. R. ROACH

061719-070 (F) 5-1/2'' SG

031204-001 ?? CRAWLER ?? RONSON

091420-001 FORKLIFT ? ROSKO

061719-070 (F) 5-1/2'' SG

081514-001 F 190 CULT. 1/12 SANWA

090803-062 CUB 1/12 SAUNDERS-SWADER
090803-059 CUB 1/12 AFINSON

061719-083B MAJOR ROADLESS 1/32 SCALEDOWN

130106-026 165 1/20 SCALE CRAFT

061719-083A MAJOR 1/32 SCALEDOWN

130106-064 165 1/20 SCALE CRAFT

010300-034B B 1/32 BRIAN PARKS
010300-034A B 1/32 BRIAN PARKS

061719-083C MAJOR ROADLESS/HALFTRACK 1/32 SCALEDOWN

061719-083D MAJOR DOE TRIPLE D 1/32 SCALEDOWN

090803-114A A 1/32 SCALEDOWN
090803-114B B 1/32 SCALEDOWN

011600-001B 8000 1/16 SCALE MODELS

201801-001 135 1/32 SCALEDOWN

030119-044 1930 (CC) 1/16 SCALE MODELS

010300-033 (WC) 1/12 SCALE MODELS

010300-048 RC 1/16 SCALE MODELS

030119-048 20-40 1/16 SCALE MODELS

030119-068B 15-45 1/16 SCALE MODELS

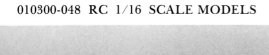

031503-007 70 1/16 SCALE MODELS

061719-066 ALL AROUND 1/16 SCALE MODELS

060901-022 44-23 1/16 SCALE MODELS

080519-002A 980 1/16 SCALE MODELS

061718-073 4000 1/12 SCALE MODELS

080519-002B 980DT 1/16 SCALE MODELS

080519-002C 1380 1/16 SCALE MODELS

090803-129B 10-10 MOGUL 1/16 SCALE MODELS

080519-002D 980DT COMM. 1/16 SCALE MODELS

090803-135 1931 REGULAR 1/16 SCALE MODELS

080519-002E 1380 COMM. 1/16 SCALE MODELS

090803-129A 8-16 MOGUL 1/16 SCALE MODELS

090803-144 F-20 1/16 SCALE MODELS

090803-154A FARMTOY (806) 1/16 SCALE MODELS

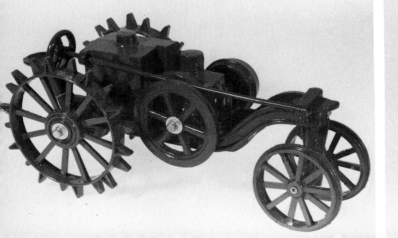

090803-154B FARMTOY (1206) 1/16 SCALE MODELS

151209-020 ROW CROP 80 1/16 SCALE MODELS

090803-154A FARMTOY (806) 1/16 SCALE MODELS

100400-086 (GP) 1/16 SCALE MODELS

090803-159 (10-20) 1/16 SCALE MODELS

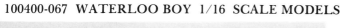

100400-067 WATERLOO BOY 1/16 SCALE MODELS

100400-100B GP ROW CROP 1/16 SCALE MODELS
100400-100B Decals added.

100400-094B (G) 1/16 SCALE MODELS

131300-023A COMFORTRACTOR 1/16 SCALE MODELS

100400-100A GP ROW CROP 1/16 SCALE MODELS

151209-023 70 1/16 SCALE MODELS

130108-033 (15-22) 1/16 SCALE MODELS

140800-002 GAS ENGINE 1/16 SCALE MODELS

190301-001 SHOW TRACTOR 1/16 SCALE MODELS

192005-008A CP-1400 PANTHER 1000 1/64 SCALE MODELS

182113-004 OIL PULL (16-30) 1/16 SCALE MODELS

192005-005A CP-1400 PANTHER 1000 1/32 SCALE MODELS

192005-011A PANTHER IV FIRST EDITION 1/32 SCALE MODELS

192005-011B PANTHER IV 1/32 SCALE MODELS

192005-005A CP-1400 PANTHER 1000 1/32 SCALE MODELS

192005-004A CP-1400 PANTHER 1000 1/16 SCALE MODELS

230112-002 (20-30) 1/16 SCALE MODELS

220518-005 825 1/16 SCALE MODELS

220518-003A 1150 1/32 SCALE MODELS

200920-001 OIL TRACTOR 1/16 SCALE MODELS

220518-004A 256 FIRST EDITION 1/32 SCALE MODELS

220518-004C 276 1/32 SCALE MODELS

220518-006A 836 1/32 SCALE MODELS

230809-001B 2-155 1/16 SCALE MODELS

230809-001C 2-135 FWA 1/16 SCALE MODELS

230809-011 (2-135) ??

230809-001I 2-155 1/16 SCALE MODELS

230809-010 700 1/16 SCALE MODELS

230809-004 4-210 1/16 SCALE MODELS

230809-009A 2-135 FIRST EDITION 1/64 SCALE MODELS

230809-006 (JUNIOR) 1/25 SCALE MODELS

230809-001H 2-135 FWA 1/16 SCALE MODELS

230809-008B 4-270 1/16 SCALE MODELS

230809-002B 2-180 1/16 SCALE MODELS

230809-007 4-225 1/16 SCALE MODELS

230809-003B 2-35 ISEKI 1/25 SCALE MODELS

230809-005A 4-175 1/16 SCALE MODELS

230809-003A 2-35 ISEKI 1/25 SCALE MODELS

230809-001A   230809-001G 2-135 1/16 SCALE MODELS

191513-001 20 1/20 SEVITA

030118-001C   JUMBO (88.4) -001B 88.4 1/24 SCAME—GIODI

191513-002 35 1/20 SEVITA

030118-001C JUMBO (88.4) 1/24 SCAME—GIODI

040521-002 60PS 1/90 SCHUCO

191513-003 40 1/20 SEVITA

040200-009 30 TD CRAWLER 1/16 SHACKLETON—

CHAD VALLEY

031521-002 SUPER 4-754 1/12 DAVID SHARP

131300-020B A4T1600 1/16 DAVID SHARP

061718-085 FW-60 1/16 DAVID SHARP

131300-021 G-1350 1/16 DAVID SHARP

090803-142 4386 & 4786 1/16 DAVID SHARP

100400-103 8010 1/16 DAVID SHARP

151209-021A 2655 1/16 DAVID SHARP

190805-001 SD-3 1/16 SHEPPARD MFG. CO.

151209-022 2155 1/16 DAVID SHARP

230809-012 4-180 1/16 DAVID SHARP

192005-012 PANTHER SERIES II 1/16 DAVID SHARP

120114-003 R5000N SPECIAL 1/15 SIACA

220518-007 935 1/16 DAVID SHARP

151209-025 770 1/16 BEN SIEGEL

021200-001 40 1/16 BEN SIEGEL

100400-160 435 1/16 SIGOMEC

030119-050A 300 1/16 BEN SIEGEL

100400-161 730 1/16 SIGOMEC

190912-001 ?? 1/16 BEN SIEGEL

100400-162 (3010) 1/16 SIGOMEC

100400-163 ?? 1/16 SIGOMEC

100400-163 ?? 1/16 SIGOMEC

100400-161 730 1/16 ?

040521-024A DEUTZ FAHR DX 86 1/32 SIKU

100400-162 (3010) 1/16 SIGOMEC

040521-024B DEUTZ FAHR DX 4-70 1/32 SIKU

060108-003 ?? 1/60 SIKU

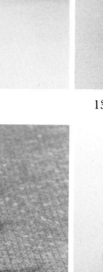

130106-054A 284-S 1/32 SIKU

060901-006A 40 CA 1/60 SIKU

130106-054B 284-S 1/32 SIKU

060901-006B 40 CA 1/60 SIKU

130106-054C 284-S 1/32 SIKU

080114-004A ROBUST 900 1/50 SIKU

130502-010 MB TRAC 800 1/32 SIKU

040200-001 (25-D) 1/16 DENZIL SKINNER

090803-023 (M) 6-1/2'' SLIK TOYS

142106-001 UNIVERSAL (M-IV) 1/16 DENZIL SKINNER

130108-017A 44 1/16 SLIK TOYS

142106-001 UNIVERSAL (M-IV) 1/16 DENZIL SKINNER
142106-004A -004B (UNIVERSAL) 1/48 DENZIL SKINNER

130108-017B 44 1/16 SLIK TOYS
131300-005 (UB) 1/16 SLIK TOYS

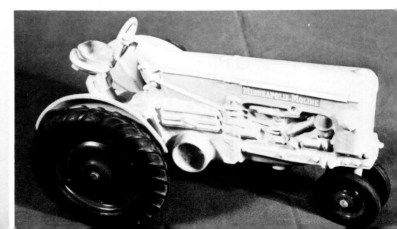

131300-006 (R) 1/32 SLIK TOYS

131300-009 (R) 1/32 LINCOLN SPECIALITIES

131300-007 (4 STAR) 1/32 SLIK TOYS

131300-008 445 1/32 SLIK TOYS

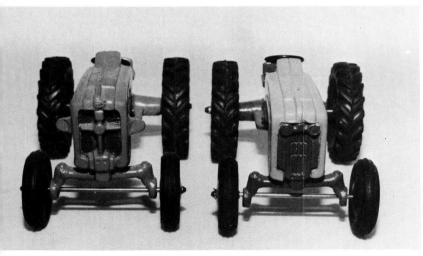

131300-007 (4 STAR) 1/32 SLIK TOYS
131300-008 445 1/32 SLIK TOYS
131300-010A (R) 1/16 SLIK TOYS

151209-005A 77 1/16 SLIK TOYS

151209-004A 70 1/16 SLIK TOYS

151209-005B 77 ROW CROP 1/16 SLIK TOYS

151209-007 SUPER 77 1/16 SLIK TOYS

151209-006A 77 R.C. DIESEL POWER 1/16 SLIK TOYS

151209-011A OC-6 1/16 SLIK TOYS

151209-008 SUPER 55 1/12 SLIK TOYS

020401-001 WHEEL TRACTOR 1/32 SOLIDO

151209-010 880 1/16 SLIK TOYS

151209-009 (880) 1/32 SLIK TOYS

090803-073 HYDRAULIC EXCAVATOR 1/55 SOLIDO

180514-008A 651-4 1/32 SOLIDO

180514-008B 651-4 1/32 SOLIDO

061719-021 (F) 1/16 CHARLES SOUHRADA

This is currently produced by Pioneer Tractor Works.

010300-011A (D-SER) 1/25 STROMBECKER

010300-011B (D-SER) 1/25 STROMBECKER

010300-011A Grill detail, -011B Variation illustration.

120126-012 ELIBULLDOG
1/43 STEAM & TRUCK

120126-013A BULLDOG 3506

120126-013B BULLDOG

060514-018 FAVORIT FW—140 1/25 STRENCO

130502-008 UNIMOG (411) 1/25 TCO

151209-030 70   STURDY STUFF TOYS

250114-001 TB-20 7'' T-N

021512-001 ?? 20 cm SUEDE PENDING

100400-079A 440 CYCLONE SNOWMOBILE 1/12
SUTTLE

060901-033 640 1/25 TTF

061719-040 MAJOR 1/43 TAISEIYA

060518-006A (30) 1/43 TEKNO

061719-052 (DEXTA) 1/24 TOMTE

090803-067 (414) 1/32 TEKNO

061719-052 (DEXTA) 1/24 TOMTE

090803-068 (574) 1/32 TEKNO

061718-061 5000 SUPER MAJOR 3'' TIMPO

061719-085 MAJOR 1/43 TIMPO

061719-047 SUPER MAJOR 1/43 TOMTE

130106-047 (1155) 1/70 TOMY

030119-020 CRAWLER 1/16 TOMY ?

061718-011 (NAA JUBILEE) 1/32 TOOTSIETOY

030119-018 580 CK 1/16 TOMY

151209-018 (1955) 1/43 TOOTSIETOY

031201-003 741 HYDROSTATIC 1/24 TONKA

061718-010A (8N) 1/32 TOOTSIETOY

061718-010B (8N) 1/32 TOOTSIETOY

060518-004 (30) 1/13 TOPPING MODELS

061719-064 (E27N) 1/250 TRAFFICAST

061719-065 (MAJOR) 1/500 TRAFFICAST

060518-018 TE-20 1/20 TRITON (LEGO)

010300-045 HD-20 1/25 TRIANG MINIC

040521-004 DM-55 1/25 TROL

202107-001 TYPE 37A 6'' TRIANG

130106-016 65 1/43 TRIANG SPOT-ON

100400-082C (820) 1/16 ELDON TRUMM

100400-083A (830) DIESEL, -083B (830) DIESEL INDUSTRIAL 1/16 ELDON TRUMM

100400-082D (820) INDUSTRIAL 1/16 ELDON TRUMM   090803-176B (M) 1/16 CARTER TRU-SCALE

100400-084 (R) DIESEL 1/16 ELDON TRUMM   (560) 1/16 TRU-SCALE

020907-001 360/30 1/16 TRUMM—DEBAILLIE   890 1/16 TRU-SCALE

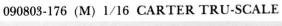

090803-176 (M) 1/16 CARTER TRU-SCALE   891 1/16 TRU-SCALE

090803-173 C 1/16 TUDOR ROSE

090803-065B (240) 1/20 UNIVERSAL-CORDEG

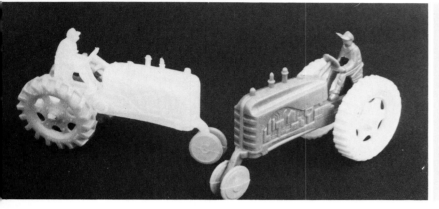

130108-034 (44) 1/30 TUDOR ROSE

030119-049 CROSSMOUNT 12-20 1/76 VM MINIATURES

061718-048 (9N) 1/12 TOY & TACKLE

192005-001 COUGAR II 1/12 VALLEY PATTERNS

260520-002 SUPER DIESEL 1/40 U. S. U. D.

192005-002A BEARCAT III 1/12 VALLEY PATTERNS

192005-006 PANTHER 1000 1/12 VALLEY PATTERNS

030119-001 (L) 1/16 VINDEX

061719-016 (F) 1/43 VARNEY COPY CAT

230112-001 (20-30) 1/25 VINDEX

220514-001 BL 1/16 VENDEUVRE

100400-001 (D) 1/16 VINDEX

020120-001 40 STEEL MULE 1/16 VINDEX

220903-001 VIGOR 1/16 VIPRODUCT VICTORY IND

061723-003 PLOWING ENGINES 1/72 W & K

131300-002 (ROW CROP Z) 1/12 WERNER WOOD & PLASTIC

100400-054 2020 1/16 WADER

061719-059 (F) 1/87 THE WHEEL WORKS

130108-003 (44) 1/87 WARDIE BJW

040521-002 (center) 60PS 1/90 SCHUCO, -003A (crawler version) ?? 1/90 SCHUCO

040521-010 ?? 1/90 WIKING

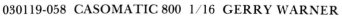

030119-058 CASOMATIC 800 1/16 GERRY WARNER

040521-011A ?? 1/90 WIKING

141500-001 NORMAL 1/90 WIKING

040521-011B ??/ROPS 1/90 WIKING

161518-003 T 1/90 WIKING

120126-005 BULLDOG 1/87 WIKING

130502-001A MB TRAC 1/90 WIKING

130502-001B MB TRAC 1/90 WIKING

161518-003 T 1/90 WIKING

100400-068A (A) 1/87 WOODLANDS SCENICS

100400-068B (A) 1/87 WOODLANDS SCENICS

021322-006 BM 1/43 YAXON

100400-069 (A) 1/87 WOODLANDS SCENICS

021322-007 BM 1/32 YAXON

060901-024 880DT-5 1/32 YAXON

061719-057 (F) 5-1/2" JOHN WRIGHT

061718-055 (NAA JUBILEE) 5-1/2" WYANDOTTE

060514-016 TURBO. FARMER 308LS 1/43 YAXON

061718-084 8210 1/43 YAXON

120114-002A, B, C 12500 or 14500 1/43 YAXON

120114-002A, B, C 12500 or 14500 1/43 YAXON

090803-130 955 1/43 YAXON

090803-130 955 1/43 YAXON

120114-002A, B, C 12500 or 14500 1/43 YAXON

120525-004 802 1/43 YAXON

130106-043 1134 1/43 YAXON

120113-001B R 1056 1/43 YAXON

130106-043A 1134 1/43 YAXON

130106-043A 1134 1/43 YAXON

130106-043 1134 1/43 YAXON

120114-002A, B, C 12500 or 14500 1/43 YAXON

120114-002A, B, C 12500 or 14500 1/43 YAXON

180514-011 145-14 TX TURBO 1/43 YAXON

190113-012 GALAXY 170 1/43 YAXON

190113-004A BUFFALO 130 1/43 FORMA-PLAST

192018-001A 8160a 1/43 YAXON

192018-001B 8160b 1/43 YAXON

190113-008 TRIDENT 130 1/43 YAXON

020401-002 ?? CRAWLER 1/32 YAXON S.A.

020412-002 (MTZ-62) 1/43 YAXON

151209-024A 550 1/16 YODER

DZ 117 1/43 YAXON-U.S.S.R.

151209-024B 550 1/16 YODER

151209-024A 550 1/16 YODER

151209-024B 550 1/16 YODER

012200-001 HUBLEY, -002 ARCADE, -003 BRUBAKER; PETERSON; WHITE; IRVIN

010300-084 UNKNOWN, 010300-085 UNKNOWN

211819-001 ?? 1/43 ZTS PLASTIK

021322-008 BM 13 cm ??

040521-005A 06 SERIES 1/32 ZISS R.W. MODELLE

010300-013A HD-14 1/20 ??

010300-001A HD-14 1/20 ??

010300-014A HD-14 1/20 ??

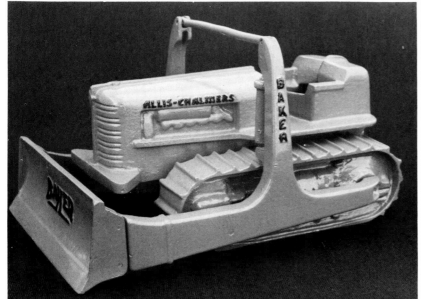
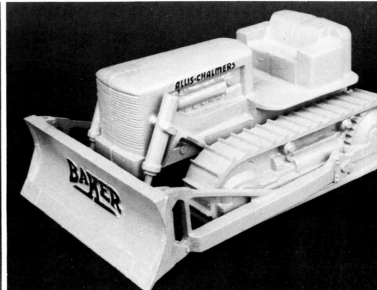

020412-001 420 (MTZ 52) 1/16 MINILUXE ?

040200-004 CROPMASTER 1/16 ?

030119-015 (2670) TK 1-1/4'' ??

042120-001 D4KB 9'' ??

030119-027 (AGRI—KING) 1/64 ?
100400-074 4430 1/64 ??
031503-009 30 1/16 ??

041714-001 2-WHEEL TRACTOR 6'' MS

061719-069 (F) / HALF TRACKS 6-7/8'' ?

080114-021 ROBUST 1/30 ??

090803-158 (M) 1/16 ??

090803-132 844 15'' ?

090803-171 (66 SERIES) 4'' ??

090803-137 (H) 1-1/2'' ?

100400-074 4430 1/64 ??

090803-174 (B-250) 12 cm ??

090803-175 844 1/43 ??

100400-153 A 1/32 ??

130106-071 175 1/16 ??

100400-166 (A) 1/16 ??

130106-072 1080 DIESEL 15'' ??

100400-168 (A) 1/64 ??

130106-038 165 1/16 ??

130108-031 PONY 820 1/20 ??

130108-047 (44) ??

131300-001 (STANDARD U) 1/12 ??

190113-009 240 DT 1/16 ??

151209-029 (70) 1/16 ??

190113-010 DA 38 DT 1/16 ??

161518-005 ?? DIESEL 12'' ??

190114-001 BEAVER 1/32 ??

182113-006 OIL PULL 1/16 ??

190913-001 GARDEN TRACTOR 1-1/4'' ??

192018-004 ?? 3-1/2'' ??

211819-002 C 335 1/20 ??

201512-001 250 ??

230820-001 ?? 1/30 ??

Miscellaneous Tractors

Code Mfger Stk# Material Size Color Country

CAST IRON

M-32 HUBLEY ? CAST IRON 4'' RED USA (Wood wheels)

M-33A ARCADE ? CAST IRON 2½'' RED USA
M-33B ARCADE ? CAST IRON 3'' RED USA

M-35 3"

M-405 IRON ART 6½" OLIVE USA

TIN

M-37 2½"

M-1 DOLE TIN 8¼" RED CI, D

M-401 OH BOY 5" RED USA

M-402 3½" RED USA

M-2 BABY TRACTOR TOY (Animated Toy) TIN 3½" GREE. USA

M-3 STRUCTO STAMPED STEEL 8½" GREEN USA (Clockwork)

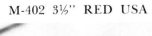

M-4 4½'' YELLOW

M-8 6½'' J

M-5 MARX "AMERICAN TRACTOR" 8½'' YELLOW CI,
USA

M-9 GESCHA 6½'' BLUE CI, D

M-6 (Military) 6''

M-7 GAMA TIN 7¼'' RED D

M-10 BETALTIN PLATE 3½" RED, YELLOW CI

M-14 TCO 7½" YELLOW D

M-11 5½"

M-13 TRACTOR 55 MARUSAN TOYS TIN 6½" RED J

M-12 3½"

M-16 MINIC TRI-ANG TIN 3¼" RED GB (Friction Drive

TIN & STAMPED STEEL

M-17B METTOY 5⅛" GB (Clockwork)

M-15 MINIC ? TIN 3¼" ? GB. (Clockwork)

M-17A METTOY #6435/77 TIN 5¼" GREEN, YELLOW CI,
GB (Mechanical Action)

M-20 K "NEW" TIN 5½" J ("Pioneer" on rear of trailer)

M-18 METTOY 9½" BLUE CI, GB

M-21 JOUSTRA TIN 8¼" RED, YELLOW F (Clockwork,
steerable, sparking exhaust)

M-19 LBZ 6" BLUE D

M-22 CKO 3½" GRAY CI, D

M-23 CORTLAND TIN 6¾'' RED CI, USA

M-26 ARNOLD 7'' D (Battery)

M-24 CORTLAND 5½'' RED USA

M-43 GAMA 3'' D

M-25 MARX ? TIN 14'' RED USA

M-52 DOLECEK-DOLTRAK STAMPED STEEL 11½'' REI
USA (Ratchet lever type wind-up)

M-44 GAMA TIN 3¾" RED, BLUE CI, D

M-109 DCGM #389 TIN 3½" GREEN 1972 D (Friction drive)

M-53 GAMA & GAMA "FORD" 4½" RED CI, D

M-110 GESCHA TIN 7½" BLUE, RED D (Wind-up, plate on front reads: ERCO 1950a)

M-103 TM 5½"

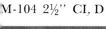

M-104 2½" CI, D

M-111 METTOY TIN 7¾" GREEN, GRAY (Steerable, "Mettoy" on front radiator. Wheels made of tin, clockwork)

M-112 SUPER 6 JUNIOR TIN 8¼" YELLOW USA (Ratchet drive)

M-113 TIN 5⅜'' BLUE P (Clockwork)

M-118 TIN 12¼'' RED, CREAM USSR

M-116 KNIB TIN 7¼'' BLUE, ORANGE 1979 (Wind-up)

M-117 ARNOLD #7350 TIN 7'' RED D (Battery, steerable, rubber tires)

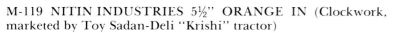

M-115 BIG G HEAVY GAUGE STEEL 11'' w/blade YELLOW USA (Rubber wheel used as steering wheel, "SLIK" type rear wheels)

M-119 NITIN INDUSTRIES 5½'' ORANGE IN (Clockwork, marketed by Toy Sadan-Deli "Krishi" tractor)

M-120 1/20 GREEN USSR

M-125 TIN & PLASTIC 4⅜'' RED, WHITE USSR

M-122 GESCHA #717 TIN 7¾'' BLUE, RED D (Battery powered)

M-130 TIN 6¾'' RED (Rubber tires marked ''MK'', Steerable, plated front grill)

M-123 SOMMAVILLA TIN 10½'' RED (Steerable, electric)

M-129 GAMA #176/4 TIN 5¼'' D (Clockwork)

M-124 FARM TRACTOR by SSS QUALITY TOYS TIN 6¾'' RED J (Friction powered, rubber tires)

M-128 GABRYDA 4¾'' D

M-131 TIN Clockwork.

M-132 ? ? TIN 3¼'' RED ?

M-134 GESCHA 1/32 GREEN D

M-137 ARNOLD #11050 TIN 11'' w/blade D (Battery powered, remote controlled with cable)

M-135 TIN 4'' long RED (Black rubber wheels)

M-138 ARNOLD #820 11cm YELLOW CI, D

M-136 ТРАКТОР TIN, PLASTIC 3¾'' RED, GREEN USSR

M-139A #701 TIN 7'' RED C ("The East is Red" Tractor, Battery operated)

M-140 MMP PLASTIC & TIN 8¼" RED BLUE (Also other colors)

M-144 KINGSBURY PRESSED STEEL 7¼" ARMY GREEN, GOLD TRIM USA (Wind-up, platform is missing)

M-139B (Variation of 139A)

M-142 MS 1/12 ORANGE DDR (Battery)

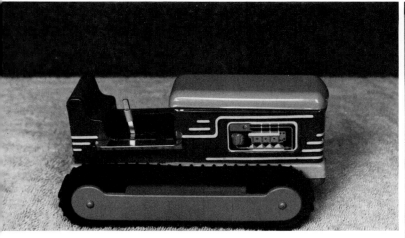

M-141 (Electric)

M-145 GAMA 7¼" CI, D

M-143 #1200 TIN 9¾" GRAY, RED J (Battery powered, pistons move up and down)

M-148 KDN TIN 7½" Variety of colors CZ (Clockwork)

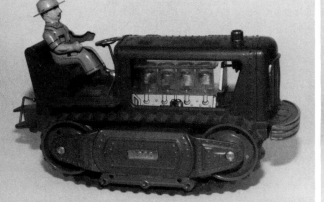

M-149 GESCHA 5¾" BLUE D

M-154 ? ? SLUSHMOLD 2¾" RED ?

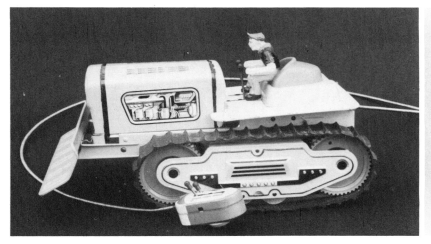

M-150 ARNOLD TIN 11" w/blade D (Battery)

M-155 TIN Clockwork

M-151 BOB'S TRACTOR by MODERN TOY #3578 TIN 1955 J
(Flashing engine, detonating engine sound, shaking drivers seat,
battery operated)
M-156 PRESSED STEEL 5½" RED ? (Wind-up)

M-153 STRUCTO STEEL 8¼" GREEN, RED USA
(Clockwork)

M-157 GEK ? TIN 6⅞" GREEN D (Wind-up)

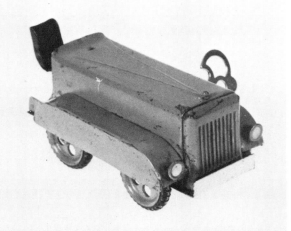

M-158 GAMA TIN 5'' RED, YELLOW D (CLockwork, parking exhaust)

M-302 UNITRAK (Anker and/or Piko PLASTIC 6'' BLUE, RED DDN (Clicker on front to make engine sound)

M-159 SOMMAVILLA TIN 27cm ? ?

M-341 TITAN TIN, PLASTIC 6'' ASSORTED COLORS 1979 CZ (Remote control, "Titan" on front of radiator)

DIECAST & SLUSHMOLD

M-301 URALEC TIN 9½'' YELLOW USSR (Battery powered)

M-30 3½''

M-31 FUN HO ? CAST ALUM. 4'' GREEN NZ (White plastic wheels, tractor resembles Ford 8N)

M-40 2''

M-38A CHARBENS 3½'' GB

M-41 TOOTSIETOY 3'' USA

M-38B CHARBENS 3½'' GB

M-39 2''

M-42 TOOTSIETOY 3½'' RED USA

M-45A BANNER ? PLASTIC 3'' GRAY ?
M-45B ? ? SLUSHMOLD 2½'' ? ?

M-46 6½"

M-51 LANSING DIE CAST 11" RED USA (With and without driver, with and without loader)

M-48 CHARBENS ? DIE CAST 3½" ? GB

M-54 M&L 5¼" RED CI

M-49 CHARBENS METAL 4" ORANGE, SILVER GB

M-56A HUBLEY DIE CAST 5½" RED USA (Cast iron driver)

M-50 METTOY GB

M-55 LEAD 1⅝" RED

M-56A HUBLEY DIE CAST 7¾'' RED USA (Iron driver, solid
rubber wheels)
M-56B HUBLEY DIE CAST 6¾'' RED USA

M-61 SLIK TOY #9890 8'' USA

M-59 LEE TOYS 3⅜'' RED USA

M-69 TOOTSIETOY DIE CAST 2'' RED USA (Yellow near
rims with rubber tires)

M-57 HUBLEY ? DIE CAST RED USA

M-60 LEE TOYS 6¼'' RED USA

M-58 TRU-SCALE DIE CAST 8'' YELLOW USA (Tin wheels)

M-70 TOOTSIETOY 3½'' USA

M-95 GS 5" USA

M-83A TOMICA BAGGAGE TR. 1/50 J
M-83B TOMICA BAGGAGE TR. 1/119 J

M-74 FUN HO 9" NZ

M-75 FUN HO 6" NZ

M-77 FUN HO 2" NZ

M-90 "CATERPILLAR ON THE WORLD" PLASTER 6"
USA

M-100 CRESCENT #1822 4" ORANGE GB

M-101A CRESCENT SLUSHMOLD ? ? GB

M-101B CHARBENS

M-355

M-105 SLUSHMOLD 2¾"

M-356 ? ? SLUSHMOLD 3" ORANGE ? (Resembles Ford 8N or MF35)

M-106 FUN HO 3¾", 3", 2" NZ

M-121 KONDOR DIE CAST 3½" ORANGE D (Clockwork, casting number is 8194/48)

M-363 BENBROS #11 DIE CAST 1/87 GB

M-403 HUBLEY 11" RED, SILVER USA

M-362A JO-HILLICO 3'' GB
M-362B

M-408 DURAVIT ORANGE AR

M-404 HILLER RED USA

M-409 DURAVIT #85 CAST METAL 10½'' GREEN AR
(Steerable)

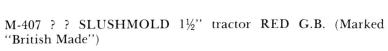

M-406 6½'' MAROON, SILVER

M-407 ? ? SLUSHMOLD 1½'' tractor RED G.B. (Marked ''British Made'')

M-410 SCALE MODELS DIE CAST ? CHROME USA (Same mold as Hubley Special Anniversary issue: Jan. 4, 1983)

M-411 CAST IRON 1/12

M-412 MEERS & SON 6" SILVER GB

M-414 STROMBECKER

M-414A TRU-SCALE

M-414B STROMBECKER 504

M-413 BARCLAY 3¼" USA

M-415 W&T ROWLAND MINIATURES 1/87" GB

M-416 ? #9764 DIE CAST 3" ANTIQUE BRONZE FINISH
HONG KONG (Pencil sharpener)

M-417 CARTER TRU-SCALE DIE CAST 1/16 scale YELLOW
USA (Front wheel assist drive, tilt away seat)

M-419 1/87 SIKU

M-63 THOMAS HOVE PLASTIC 5'' YELLOW AUS.
(Available in assorted colors)

M-418 CHARBENS ? SLUSHMOLD 3½'' ORANGE, LIGHT
BLUE WHEELS G.B.

M-47 LEAD 1⅜'' RED

M-52 1/16 SCALE MODELS

PLASTIC

M-62 8'' RED, YELLOW

M-64 DBCM ''BIG'' #910 PLASTIC 12½'' GREEN D
(Steerable)

M-65A GAY TOYS 7½'' RED, SILVER USA

M-65B GAY TOYS PLASTIC 7½'' BLUE, WHITE USA

M-73 NASTA 3'' GREEN 1976 USA (Friction powered, made in Hong Kong)

M-66 MARX PLASTIC 4⅜'' RED J

M-76 FUN HO 4'' NZ

M-71 TOMY-MATIC 340 9'' YELLOW J

M-72 BUDDY-L 3½'' USA

M-78A LESNEY MOD 4'' GB
M-78B LESNEY MOD 2'' GB

M-86 BANNER PLASTIC 3'' RED, YELLOW USA

M-96 MULITPLE PRODUCTS 5½'' USA

M-126 YELLOW USSR

M-79 9''

M-98 PLASTO ? PLASTIC 9½'' GRAY FIN

M-89 PAYA PLASTIC 12'' ORANGE E (Remote controlled)

M-303 #1042 PLASTIC 11½'' GREEN, YELLOW, RED USSR

M-97 STEHO PLASTIC 6'' RED D

M-133 TONKA TIN, PLASTIC 4⅜'' GREEN, YELLOW USA

M-304 SIKU 1/32'' YELLOW, GREEN 1960 D

M-310 PLASTIC 11'' BLUE USSR (Steerable)

M-305 ? PLASTIC 3½'' YELLOW ? ? (Red steering wheel, black wheels)

M-308 JEAN 1/50'' MILITARY GREEN D

M-306 STEIFF ORANGE D

M-309 TA TM CHEMA PLASTIC 6¾'' RED, WHITE CZ

M-311A ARTHUR HAMMER PLASTIC 7¼'' GREEN, RED, YELLOW D (Hood lifts)

M-312 GAY TOYS 13'' USA

M-307 RED

M-317

11B ARTHUR HAMMER PLASTIC BLUE, WHITE (Similar
311A with addition of cab)

M-316 GOLIATH PLASTIC

M-314

M-318 "REX" PLASTIC 11½" YELLOW, BLACK D (Battery
perated)

M-315 ? ? PLASTIC 6½" BLUE & YELLOW ("Farmulus" name
on side of hood near front)

M-313 IRWIN 5½" RED, GRAY USA

M-319 PLASTIC 7'' RED, BLACK

M-324 SOFT PLASTIC 6'' YELLOW USA (A ''squeeze toy'')

M-25 M-323 Plastic

M-325 PLASTO #1915 PLASTIC 9'' YELLOW & RED FIN

M-321 GOSO 5½'' RED, YELLOW D
M-320 BARR RUBBER PRODUCTS COMPANY ? PLASTIC
4⅜'' COLOR VARIATIONS USA (''Ohio'' name on side of
tractor. Made in Sandusky, Ohio)

M-329 PLASTIC 5½'' COLOR VARIATIONS 1979 CZ

M-326 SANSON PLASTIC 6'' ORANGE & YELLOW I

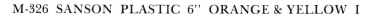

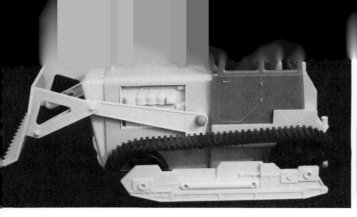

M-330 PIKO "UNITRAK" PLASTIC 8" with blade BLUE, RED, GRAY CZ

M-335 PLASTIC 6" BLUE, RED, YELLOW

M-327 CHEERIO TOYS PLASTIC 3¼" RED, YELLOW CDN

M-333 DUBENA CESKY DUB PLASTIC 6" RED & BLUE CZ
(Pull toy, trailer included)

M-332 MARX PLASTIC 5½" ORANGE HK (Battery operated)

M-337 PLASTIC 10¼" GRAY, YELLOW, RED (Battery operated)

M-336 PLASTIC 6½" YELLOW, WHITE

M-340 PLASTIC 3⅝" ORANGE CZ

M-342 PLASTIC 6'' BLUE J (Battery controlled)

M-347 ERTL GULLY WHUMPER PLASTIC 5'' YELLLOW USA

M-338

M-346 PLASTIC 3½'' BLUE, RED, YELLLOW, WHITE BLACK USSR (Modified racing type model)

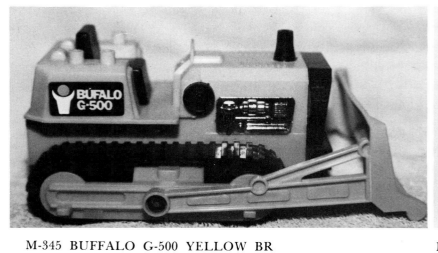

M-345 BUFFALO G-500 YELLOW BR

M-344 JEAN PLASTIC 4'' long BLUE, SILVER, YELLOW D (Steerable)

M-348 ERTL CLOD HOPPER PLASTIC 5'' RED USA

M-349 CANE GIOCATTOLI 1/32'' I

M-351 PLASTIC 3¼" RED (Bk wheels)

M-357A NOREDA PLASTIC 2¾" GREEN, ORANGE F (A "Renault" type model of the wheel and crawler versions of the tractor)

M-352

M-357B NOREDA PLASTIC 2¾" GREEN, ORANGE F (A "Renault" type tractor)

M-353 PLASTIC 4½" ORANGE USSR (Clockwork)

M-358 BRUDER PLASTIC 1¾" 2WD, 2" 4WD RED, GREEN, YELLOW D

M-354 E. ALLAN BROOKER PLAYMOBILE #3500 PLASTIC 10½" GREEN 1974 NZ

M-359 PLASTIC 2¼" YELLOW, GREEN

M-360 PLASTIC 2¼" RED, WHITE

M-366 CO-MA #5156 PLASTIC 13" ORANGE-BROWN YELLOW-GREEN I (Tilting hood)

M-361 PLASTIC 2⅛" YELLOW

M-367 EMPIRE ? PLASTIC 14½" ORANGE U (Resembles an "D" Series tractor)

M-364 FLEETLINE N-GAUGE (Kit #N-95)

M-365 COLECO ? ? ? RED USA

M-368 E. ALLAN BOOKER LTD. PLASTIC 3" RED NZ (Friction motor, front mounted leader, resembles MF135 tractor Made in Hong Kong)

M-369 LARAMI CORP #7000-3 PLASTIC 3' GREEN USA (Old Mac Donald's "Old Mac" tractor modified V-8)

M-370 ? ? PLASTIC 2⅛" ORANGE ?

M-374 ? ? PLASTIC 3" RED ? (Resembles "Porsche" tractor)

M-371 ? ? PLASTIC 3" ORANGE ? (Resembles Fordson Dexta)

M-375 GAMA #8015/8016 PLASTIC 11" RED, GRAY D (Auto type steering, battery operated-batteries stored under tilting hood. "Favorit" model tractor)

M-372 ? ? PLASTIC 1¼" COLOR VARIATIONS ?

M-373 ? ? PLASTIC 1⅜" RED ?

M-802 FARM-O-CRAFT

M-30 M-803 1/64

M-27 ? ? WOOD 4" ? ?

M-97 WOODEN TOYS WOOD 12" USA ("Farm Help" imprinted on top of hood)

M-28 ? ? WOOD 10½" ? ?

M-502 WOOD 10⅞" RED, YELLOW TRIM

M-29 CHAD VALLEY 7" GB

M-87 BRIO 4½" S

M-503 CO-OP WOOD USA

M-506 WOOD 4" BLUE, GREEN, YELLOW, RED CZ (A pull-toy)

M-504 T. POLI 1/16 1950 I

M-508 MINI-TOY WOOD & METAL 3⅛'' RED & YELLOW

M-505 WOOD 3½'' GREEN, YELLOW, RED

M-510 209 UNIVERSAL USA

M-507A WOOD 4¼''
M-507B WOOD 7¼'' (Larger tractor is steerable, pull-toy)

M-509 PETER-MAR USA

M-511 TRAKTORIUS WOOD 7" NATURAL FINISH 1979
USSR ("CCCP" stamped on bottom)

M-514 DEE BROS ? WOOD 11" RED USA (Farm toy D)

COMIC

M-512 ENOS 1922

M-84 MARX DONALD DUCK 3½" HK

M-513 ? WOOD 13" RED ? ? ("Dunlop" marked rubber tires,
steerable)

M-85 SUN RUBBER MICKEY MOUSE 4" RED USA

M-200 ERTL OL' McSMURF 2¾'' ORANGE 1982 USA

M-801 COMPOSITION 8½'' RED USA (Has ''clicker'' for engine noise, black wheels)

M-201 SUN RUBBER DONALD DUCK 4'' RED CDN

M-500 HUBLEY CEREAL PRODUCT 3¾'' long RED USA (WWII era, referred to as an edible tractor—made from cereal)

GLASS

M-80 AVON ''THE HARVESTER'' GLASS 6'' AMBER USA (After shave lotion bottle)

M-99 HUBLEY #125 HUBLIOD MATERIAL 5½'' RED USA (A pull-toy, driver's head is cast iron)

M-82 AUBRUBBER 5" RED or RED & SILVER USA

M-81 AUBRUBBER RUBBER 3⅞" USA

Discriminating Youngsters Unanimously Endorse the Distinctive NEW JOHN DEERE TRACTORCYCLE

"I've Ridden It... and I'm Convinced"

Of course, the young "tractor connoisseur" is a bit prejudiced. His dad's been talking a lot lately about the exclusive new features of the brand-new John Deere Tractors. Fact is, Dad and his John Deere dealer have been putting their heads together on a new tractor. Now that Junior's sold on John Deere . . . looks like Dad's a *sure bet*.

You'll promote your tractor and implement line . . . build good will among customers and prospects by displaying and selling the New John Deere Tractorcycle. You can add a handy new profit, too, especially with Christmas coming up.

Youngsters Love Them

Youngsters will love the new John Deere Tractorcycle because it's "up-to-the-minute" in the latest John Deere Tractor features. To kids, it's . . . "just like Dad's."

The new John Deere Tractorcycle highlights the spanking new John Deere Tractor two-tone color scheme . . . distinctive new front-end trademark . . . "power steering" steering wheel . . . adjustable seat. It's made of heavy cast aluminum, and boasts sturdy chain drive, heavy, solid rubber tires, and two coats of long-lasting, gleaming enamel. Big-capacity two-wheel trailer and miniature John Deere Umbrella are popular "extras."

Be sure and stock Tractorcycles and other John Deere Farm Toys. They'll provide promotional, advertising, and good-will value that you just can't buy elsewhere.

THE ESKA COMPANY
100 West Second Street, Dubuque, Iowa

Pedal Tractors

6

INTRODUCTION

Pedal Tractors are an action toy made for smaller children to "play" farming—just like dad.

Some of you may recall earlier years when you used whatever available materials, such as wood from orange crates, discarded buggy or wagon wheels, and fashioned some sort of "tractor" out of these materials. Maybe you were more fortunate to have a grandparent or dad and mom who purchased one of those nice new store bought models from your local machinery dealer for you. These were the most desirable pedal tractors because they "looked just like the real" tractor. Perhaps this is why collecting pedal tractors has become popular. They do represent the real tractor in color and design.

In recent years, collectors of pedal tractors have grown in number, more beginning every day. Toy shows and auctions will find pedal tractors displayed and for sale. It has grown to the point where individuals have made a sideline business of procuring, manufacturing, and selling replacement parts and decals. This means that collectors can restore these "greatly used" collectables in as-near-original condition as possible.

HISTORY

The pedal tractor was a more recent development than the smaller toy versions of farm tractors.

According to Eldon Trumm, author of the book *Pedal Tractors*, a Dubuque, Iowa, man by the name of Harold Heller made a pedal tractor out of wood. He then showed it to the John Deere and International Harvester headquarters. This resulted in these companies accepting the new idea of this kind of child's toy. Mr. Heller formed a company called "Tractoy" which later merged with the "Eska" company of Dubuque in 1949. The Eska Company produced a wide variety of name-brand pedal tractors. These are the most sought after models by collectors today.

In 1960, Eska ended its pedal tractor production. Thus, the ERTL Company of Dyersville, Iowa, began making and marketing pedal tractors—the 4020 (20) John Deere being the first.

MAKING THE PEDAL TRACTORS

The production method of making pedal tractors has also changed over the years. Early models were sand cast in aluminum. Today, steel molds are used to receive the 1600° F liquid aluminum alloy.

After the molten aluminum is poured into the mold while on its side, it is then hydraulically lifted to vertical position for 60-70 seconds, the mold is then opened and the cast removed. Approximately 26-30 castings of a tractor half are made every hour, each weighting 8-10 pounds. Castings are then cleaned of flashing, holes are pierced, and rough edges are ground and sanded. Assembly begins with installation of front post and steering shaft. The two halves are then bolted together and hard-to-reach surfaces are hand filled. The sanding booth finds operators using disc sanders to all surfaces to insure a "good fit."

Assemblies are then ready to be placed on a 400 foot overhead conveyor to be washed and oven dried. A coat of iron phosphate is added to aid the adherence of the electro-static paint. After 5-7 minutes in the bake oven, the paint is dry. Final asembly consists of installing front wheels. The main body, plus remaining parts—steering wheel, seat, rear wheels, and decal set—are then packaged for shipping.

One hundred ninety (190) riding cycles are completed each hour. One thousand to twelve hundred (1000-1200) are produced each week, eleven months out of the year.

Where do they all go??? And you said you were having a hard time finding pedal tractors?

PEDAL POWER

A more recent development related to pedal tractors has been the emergence of pedal pulling contests. Again, the little folks have their chance at trying their strength and skill in "out-pulling" their competition—just like the big boys!

Pedal tractor pulls have been taking place all across the country. They appear at local community fairs, county fairs, and other rural festivals. Interest in pedal pulling is evident in the long line of little people, with nervous anticipation, awaiting to climb aboard and show their stuff. Oh, yes, the girls are tough competitors, too.

The event is run much the same as the real National Tractor Pullers Association (N.T.P.A.) pulls. A sled with movable weight moves gradually forward placing a greater pressure on the sliding pan, thus, more pull resistance to slow or stop the contestant. The pull is completed when the driver spins out or can no longer turn the pedals.

"Classes" vary from one location to another such as by age groups, girls groups, contestant weight, etc.; also, stock and superstock or modified tractor classes. A stock pedal tractor is one with no mechanical or design changes—just as it came from the manufacturer. Modified units may have been made into a four wheel drive, geared down with sprocket size changes, longer pedal crank, tire and wheel size changes. All these add up to improved pulling performance.

Usually a pull will be sponsored by a local promoter. The pulling equipment, sled, and pedal tractors are furnished by someone that makes a business of providing the machines. Oh, yes, there are written rules governing many of the contests, including state level elimination contests.

FOREIGN MANUFACTURERS

Thus far, our attention has been on pedal tractors made in the U.S.A. Other countries, especially on the continent of Europe, produce very nicely detailed promotional models of "ride-on" tractors.

The "Rolly Toys" brand of pedal tractors are made in West Germany by the Franz Schneider GMBH and Company. Of the foreign manufacturers, this company is perhaps most known by U.S.A. collectors because of the "Deutz" line that was imported and sold through dealers. Also, the John Deere 3140 interested American collectors along with the Massey Ferguson. These are all recognized brands of "real" tractors in America. Another West German company, Martin Fuchs of Nurmberg, produced several fine ride-ons. Other non-

"Big wheels, little wheels" October 1984 calander, Deere & Company

A determined young contestant, "pouring on the coal" at a pedal tractor pull.

U.S.A. companies include: Falk of France; Ferbedo of West Germany; Ampa France of France; Sharna and Tri-Ang of England.

In reference to the foreign made pedal tractors, it should be noted that in addition to their fine detail and European styling, the main differing feature from U.S. makes is their construction of tough durable plastic. All the non-U.S.A. models in this book are of the plastic material rather than cast aluminum alloy.

The U.S.A. manufacturers include: Graphic Reproduction of Detroit; Roadmaster Corporation of Olney, Illinois; Inland Manufacturing; AMT; Bingham Manufacturing Corporation; New London Metal Processing Corporation of New London, Connecticut; Eska of Dubuque, Iowa; and The ERTL Company of Dyersville, Iowa.

USES IN SALES AND PROMOTION

The pedal tractor, being a child's toy, is looked upon by the manufacturers of the "real" models as a sales promoting idea. Their primary purpose for selling them through the dealer organization is not to just produce a toy, but to provide another avenue to get before the young public their brand-name and colors. Don't underestimate the power of molding the mind of today's youngsters to becoming "a John Deere man" in later life.

THE EARLY FORD PEDAL TRACTORS

One of the lesser known U.S. pedal tractors manufacturers is "Graphic Reproduction." Mr. H. Allen Jefferies, owner of Graphic Reproduction printing business and Carl Baier (deceased) former district manager for G.M., felt a child's pedal tractor would be a good promotional item for Ford tractor. They had a proto-type made and presented it to Lester Birger of Ford Advertising Department. He felt it was a good replica and gave Graphic Reproduction permission to proceed with production of the pedal tractors.

Mr. Jefferies' printing company was "Graphic Reproduction." Since the pedal tractor was a reproduction of the real Ford tractor, they selected the "Graphic Reproduction" name to produce and market the pedal tractors to Ford dealers. A sign, "Graphic Reproductions—Tractor Division," was made and hung in front of the leased Foundry in Three Rivers, Michigan, where castings were made.

A brochure showing the new Ford pedal tractor was printed and distributed to all Ford tractor dealers. According to Mr. Jefferies, "Ford Dealers were clamoring for them. We had an order for 300 from a dealer in Florida. By the way, it was a good unit, I gave one to a young nephew and he rode it for years and then afterwards his brothers and sisters rode it."

Even though there was excellent dealer support, the project was doomed. A ceiling price of $21.50 was established by Ford. This was less than it cost to produce the units. Therefore, Graphic Reproduction was losing money on each unit made. All of the 3000-4000 orders received were not filled. Eventually, the project was discontinued.

Two castings patterns were used in making the Ford pedal tractor. The early pattern produced some problems, so another pattern was made and used. Apparently an update in tractor model design was incorporated in the later version. Patterns have since been scrapped, according to Mr. Jefferies.

PICTURE SECTION

Following is a selection of pedal tractor pictures from the authors' collection and those of several other collectors. Where pictures were not available of pedal models, catalog illustrations were used. In many selections, a color picture is provided to assist collectors in identification and for use in restoration projects.

Model designations in paranthesis ( ) indicates the writers guesstamation as to the model it represents.

A small box is provided by each picture for collectors to use as an inventory check-off list.

In the chart listing of models, the "year" represents as nearly as possible the first year of production of the real tractors. The actual year the pedal tractors were made may be different.

Casting code numbers represent the year the design was made for the pedal tractor casting. Example: F-65; Ford 1965.

During the search for information on this section, not all desired data was discovered, therefore, as time goes on, new information will be obtained and new models added.

☐PT0103-001   (c)   Eska   34.5" long
Restored model shown. A chain tightener sprocket is located
where vertical slots are shown on rear housing.

☐PT0103-002   "CA"   Eska   38.5" long
Restored model shown. Original front wheels were metal
with hub caps like used on rear wheels. Pedals were of the
early metal end, tear drop type.

☐PT0103-003   "D-14" Eska   38" long
Cast in grill design, D-14 decal designation, 1¼" wide front
tires. Original had metal end, tear drop type pedals.

☐PT0103-004   "D-17" Eska   39" long
Smooth grill with decal, white wheels, seat, and pedals. Hub
caps were wheel color.

☐PT0103-005   "190"   Ertl   37" long
Known as the "Bar Grill" version, or the "Arrow" hood decal
version. White wheels with plated hub caps. Casting code:
A-64. Original had radial spoke design rear wheels.

☐PT0103-006b   "190 XT" Ertl   37" long
No horizonal bars in front grill.

□PT0103-007   **"200"**   **Ertl**   **37" long**
Same casting as -006b but with decal change. Push nut caps used on front plastic wheels.

□PT0103-008   **"7080"**   **Ertl**   **37" long**
Orange and maroon. Note absence of hand hold, front grill latch and vertical grooves on side of hood as used on -007. Molded plastic seat. This casting does not conform to the style of the real tractor.

□PT0103-009   **"7045"**   **Ertl**   **37" long**
Same casting as "7080". Colors changed to orange and black. Note: this body style does not conform to that of the real tractor.

□PT0103-010   **"7080"**   **Ertl**   **37" long**
Orange and black. Same as "7045" except decal change. Note: body casting does not conform to the style of the real tractor.

□PT0301-001   **(VAC)**   **Eska**   **35" long**
Flambau orange, 1" x 6" front wheels, original pedals were the metal end tear drop type. Original decal included the Case eagle on side panel directly above pedal crank. Cast seat with spring steel support was used. Hub caps were spherical shaped and painted wheel color. Chain tightener used near rear of housing.

□PT0301-002   **(400)**   **Eska**   **38.5" long**
Case Desert Sunset and orange color. Equipped with noise maker, pressed steel seat and spoke design wheel pattern. Original pedals were tear drop, metal end design. This

model has mistakenly been taken as a Farmall "M" when found incorrectly painted or no paint at all. Note: Hood decal should be placed nearer to front. (see "Eska" advertisement)

□PT0301-003   "Case-O-Matic"   Ertl   38" long
Desert Sunset and orange. Noise maker, 1½" wide front tires. Wheels of the spoke design.

□PT0301-004   "Pleasure King 30"   Ertl   39" long
Desert Sunset and orange. 1½" wide front tires. This model began using plastic pedals.

□PT0301-005   "AgriKing 1070"   Ertl   35.5" long
Desert Sunset and orange. No spoke design on wheels.

□PT0301-006   "AgriKing"   Ertl   35.5" long
Case white and Power red (new colors). Plastic front wheels with push nuts (no hub caps) were used.

□PT0301-007   "Case" (90 Series)   Ertl   35.5" long
Case white and power red. Plastic front wheels. Note: This casting does not conform to the style of the real tractor.

□PT0402-001   Model ?   Sharna   length?
This David Brown model displays the true colors as follows; Seat, grill, pedals and steering wheel- power red; wheel center, fenders and hood-David Brown Orchid white; other parts-chocolate brown. Note: Vertical exhaust stack is missing.

□PT0615-001 (900) Graphic Reproduction 40" long
Red and Gray. Front grill recessed with fine vertical grooves, steel spring seat support. Cast aluminum fenders. Steering wheel shaft installed at more slant than on -002. Note: Head lights protruding from front of hood are missing from this model. Holes in frame differ from those on -002. This model of the real tractor was made from 1955-1957.

□PT0615-002 (901) Graphic Reproduction 40" long
Gray and red with black seat. Grid-type grill design, pressed steel fenders, (some reported with cast). Hole pattern on frame differs from -001. Headlights protruding from front hood are missing on this pictured model along with round front medallion. Has cast in seat bracket. Original wheels are gray, hood luver-black. Cast in "Ford" on hood-red. Steering wheel shaft more slant than on -001. Real tractor was made in 1958-1962.

□PT0615-001
Front view of -001 showing grooved, vertical grill desisgn. Original model displayed a round Ford 900 medallion at top of grill. Recessed grill should be gray.

□PT0615-003 "Commander 6000" Ertl 38.5" long
Ford blue and gray. Steering wheel and seat are gray.

□PT0615-001
Rear view of -001 showing spring steel seat support and cast fenders. Seat painted black.

□PT0615-004 "8000" Ertl 36" long
Plastic steering wheel, seat and front wheels.

275

Within the image:
A partir de 6 ans:
Tracteur à chaîne "Ford", en plastique
moulé, direction à crémaillère.
Long. 130 cm.

175 F

□PT0615-007   (4000)   Mfg'er?   130cm long
Ford blue and white. Note: detailed luvers, raised letters and
grill detail.

□PT0615-008   "TW-20"   Falk   35" long
Plastic, horn equipped, cleated front tires. Ford blue.

□PT0615-005   "TW-20"   Ertl   36" long
Same casting as "8000". This model does not conform to the
style of the real tractor.

□PT0615-012   (4000)   Tri-Ang   length?
Plastic, with cleated front drive tires. Red seat and pedals,
dark gray wheel covers, light gray engine, white grill,
remainder Ford blue.

□PT0615-006   "7700"   Fuchs   length?
Plastic, Ford blue and white, protruding hood lights at sides
of grill.

□PT0405-001A   "5006"   Rolly Toys   35" long
Plastic, dark green with silver gray engine, orange seat, white
steering wheel.

□**PT0405-001B** "5006" Rolly Toys 35" long
Plastic, lighter green (Spring Green) than -001. Same casting as -001.

□**PT0405-005** "DX 85" Falk 22.5" long
Plastic with single front wheel, tri-cycle design.

□**PT0405-002** "4.70" Rolly Toys 31" long
Plastic, this model displays the "Fahr" name in the front grill resulting from a merger of the two companies.

□**PT0605-001** "Favorit Turbomatic" Rolly Toys 35.5" long
Plastic promotinal model of the Fendt.

□**PT0405-004A** "DX110" Rolly Toys 38.5" long
Plastic, representing the later DX series of Deutz tractors. Features fenders and front weights.

□**PT0318-001A** "D-4" New London Metal Processing Corp. 36" long w/o blade, 40" with blade
A working model of the famous "Caterpillar" crawler (track laying) tractor. Has individualized track segments; steers; blade lifts. Made of stamped steel construction with silk screened engine details. Tool box with lid on each side of seat.

□**PT0908-001** **(H)** **Eska** **33" long**
Open grill version. Cast aluminum seat, small 1" x 6" front wheels, smooth rear wheels. Larger exhaust muffler and different engine detail than -002.

□**PT0908-004** **(M)** **Eska** **38.5" long**
Larger casting with absence of muffler and air cleaner intake stack. Pressed steel seat, "noise maker" shift lever, 1½" wide front tires. Finger grip type steering wheel.

□**PT0908-002** **(H)** **Eska** **35.5" long**
Closed grill version. Steering rod support is separate casting. Referred to as the medium size Farmall. Stub on hood next to air breather used for rubber hose to simulate exhaust stack.

□**PT0908-005** **"400"** **Eska** **39" long**
Cast-in grill design. Spoke pattern wheels, tear drop, metal end pedals. Pressed steel seat.

□**PT0908-003** **(H)** **Eska** **33" long**
Another closed grill version characterized by its high steering shaft support. Note: spring steel seat support.

□**PT0908-008** **"806"** **Ertl** **36.5" long**
This model represents another major styling change of IH tractors. Noise maker shift lever is absent and new design cast aluminum seat used. Plastic pedals begun being used. 1½" wide front tires. Finger grip steering wheel.

□**PT0908-017**   **"633"**   **Rolly Toys**   **24.5" long**
Same casting as -014. Black steering wheel.

□**PT1004-001**   **(A)**   **Eska**   **33.5" long**
First of the John Deere pedal tractors. Open grill design. Spherical shaped hub caps painted yellow. Smaller, three spoked steering wheel not used on any other model. Seat design used only on this model.

□**PT1004-002**   **(60)**   **Eska**   **34" long**
Smaller of the two "60" casting variations. Steering wheels variations included the steel ring with flat cross bar as shown, and the cast aluminum medium size, three spoke design. A short length of rubber hose was supplied with this model to simulate exhaust stack. Hub caps were yellow on spoke design wheels.

□**PT1004-003**   **(60)**   **Eska**   **38" long**
Simular casting as -002, but larger in size. Finger grip steering wheel and noise maker shift lever begun beging used on this model.

□**PT1004-004**   **(620)**   **Eska**   **37.5" long**
Very simular casting to -003, except for the raised portion of casting on the front top of grill for placement of medallion decal. The main distinguishing feature of this model from the "60" is the broad yellow decal on hood sides.

□**PT1004-005a**   **"130"**   **Eska**   **38" long**
The last model representing the two cylinder John Deere tractor. This model represents the model number "630"—"730" of the real tractor. It's characterised by its smooth hood design and wide yellow band on its hood sides. Two variations exist. This version has open area between bottom of hood and top of drive shaft for radiator fan. Only one bolt

used to fasten casting together above front wheel area. Some models had dash and power steering decals. (see -005B for other variation)

□PT1004-007   "20"   Ertl   38" long
This model represents the 3020-4020 series of the real tractor. Note the absence of holes in engine compartment. PTO clutch lever is molded on left side of hood cowling. Has a dash decal showing "Power Shift" pattern. Very popular model made several years. Casting code -65.

□PT1004-005b   "130"   Eska   38" long
Nearly same as -005A except area above radiator fan drive shaft is closed in. Two bolts are used to fasten castings together above front wheel area.

□PT1004-009   (4430)   Ertl   37" long
New casting patterned after the new "30" series of early 1970's. Begun use of plastic wheels, seat, steering wheel and self-adhesive stick-on decals. Casting code 520.

□PT1004-006a   "10"   Eska   37" long
This model represents the 3010-4010 series of the totally redesigned new, six cylinder tractors. Yellow seat and wheels. Noise maker shift lever no longer used. Two variations exist. This version has the three holes in the engine area on the left side. (see -006b for other variation)

292 - JOHN DEERE TRAKTOR
cm. 92 x 43 x 59
1 pezzo
dmc. 194
kg. 9.5

□PT1004-006b   "10"   Eska   37" long
Same casting as -006a except it has four holes on left side of engine compartment.

□PT1004-001a   "3140"   Rolly Toys   40" long
Plastic model of European style John Deere features front end suitcase weights, headlights, working horn and hitch with drawpin.

□PT1306-001   (1130)   Ertl   36.5" long
First of Massey-Ferguson pedal tractors by Ertl. Plastic front wheels, steering wheel and seat. Casting code 1100.

□PT1306-004a   (188)   Rolly Toys   37.5" long
This model simular to -003 but larger in size and doesn't have fenders. Has larger wheels and cleated front tires.

□PT1306-002   (2775)   Ertl   36.5" long
Same casting as -001 except decal style change. This casting design does not correspond to that of the real tractor.

□PT1306-005   "145"   Falk   36" long
Nicely detailed European model of "Multipower" 145. Plastic, with cleated front tires. Blue fenders.

□PT1306-003   "188"   Rolly Toys   30.5" long
Plastic model of European styled M.F. tractor.

□PT1306-007   "2640"   Falk   30" long
Plastic, model displaying the later series designation "2640" Equipped with cleated front and rear tires, and fenders.

Pressed steel construction, plastic seat with back rest, large plastic rear wheels, wide front axle and chain drive.

□**PT1308-001**　　　　　(44)　　**???**　　　　**33" long**
Open grill with cast aluminum seat and unusual spring and seat support design like the "Velvet Ride" spring mounted seat used on the real Massey-Harris tractor. Decal on inside of seat reads: "Velvet Ride tractor seat for your comfort and health", along with the large red MH and Massey-Harris name. Unique features differing from other tractors include: air intake and muffler stacks, no hub caps-push nuts only; a U-joint on steering shaft, regular non-skip tooth chain and sprocket; front tire markings—SWAN 7 x 1.25 puncture proof; rear tires marked "puncture proof" 12 x 1.75.

□**PT1309-007**　　　"**Farmer Boy**"　　**Mfger?**　　**length?**
Cast aluminum hood and grill piece, with raised letter cast in, smooth rear tires, single front wheel and belt drive.

□**PT1308-002**　　　"**44 Special**"　　**Eska**　　**38.5" long**
Nicely detained aluminum casting of the Massey-Harris "44" Special as designated by its decal. Larger in size than -001. Has noise maker shift lever and tear drop metal end pedals.

□**PT1309-009** "**Tractor Junior**"　　**BMC 36" long (Junior)**
　　　　　　　　　　　　　　　　　　　40" long (Senior)
Pressed steel construction with push type pedals. Available in two sizes: "Tractor Junior" 36" long, "Tractor Senior" 40" long. The Senior version has chain drive. The models were made by "BMC" Manufacturing Corp.", Binghamton, N.Y. An interesting feature of this model is the availability of various accessories including sulky and towing cart (shown above), bulldozer snow plow, hay rake and trailer.

□ **PT1309-010a (Farmall C)   Inland Mfg' Co. 41.5" long**
Cast aluminum hood with raised letters, gear shift and throttle levers for realism. Red enamel. Sold under name of "Tractall".

□ **PT1313-001   "Tot-Tractor"   BMC   40" long**
Tubular and stamped steel construction with chain drive. Single front wheel. Painted Minneapolis-Moline Prairie Gold and trimmed in red with large red MM decal displayed on each side of hood. Note: This model does not conform to the appearance of the real MM tractor. Comparing this tractor with PT13090-009 shows this to be the same tractor as the BMC tractor except push pedals are used as were used on the BMC tractor "Senior". Two variations: *Early:* U-shaped front wheel support strap was narrow. Cotter keys used to hold wheels. *Late:* Front strap support was widened to increase strength. Push nuts used to hold on wheels.

□ **PT1313-002 "Minneapolis-Moline" BMC 41" long**
Stamped steel construction with "bullet" shaped front end. "MM Minneapolis-Moline" decal on hood. A shift lever operated the "Shuttle Shift" which consisted of two drive sprockets and chains. A shift quandrant with notches L-N-H provided for speed changes. Colors are MM Praire Gold and red. This model does not conform to the styling of the real tractor although sold through MM dealers. (above picture missing spark plugs and wires.)

□PT1512-001 "88 Diesel Power" Eska 38" long
Colorful, smaller size of the early Oliver Row Crop 88 with open style grill. Cast aluminum seat. Meadow Green body, yellow grill and red wheels and spherical hub caps.

□PT1512-002 "88" Eska 33" long
Closed grill version of -001. Red wheels.

□PT1512-003 "Super 88" Eska 39" long
Larger casting than -001 and -002. Grip type steering wheel and noise maker shift lever added. Green wheels. Pressed steel seat, spoke design in wheels, metal end tear drop pedals.

□PT1512-004 "880 Diesel" Eska 39" long
Representing the "880" series, this pedal tractor uses a decal for grill design. It is the first of Oliver tractors to begin using Clover White in the color scheme on wheels, trim and decals.

□PT1512-005 "1800" Eska 39" long
Very colorful model of the first of the 1800 series of tractors. This model had white pedals and solid cast in grill utilizing a decal for accent. This model is referred to as the "Checkerboard" version because of the hood decal design.

□PT1512-007 "1800" Ertl 39.5" long
First Ertl made Oliver. The grill was a separate thermomolded plastic piece. This model is referred to as the "Red Boarder" 1800 due to the red area around the 1800 designation. Casting code 0-63.

□PT1512-008 "1850" Ertl 39.5" long
Same as -007 except model designation change on decal.

284

334. Tracteur à remorque.
Tracteur à pédales avec sa remorque. Roue AV directrice antidérapante. L. 110 cm. . . . . F 164,00

□**PT1512-009**   **"1855"**   **Ertl**   **39.5" long**
Same as -007 and -008 except model designation and red border around model number deleted.

□**PT1805-003**   **"Renault"**   **Falk**   **110 cm long**
Plastic, tri-cycle front end, orange color.

331. Tracteur Renault.
Vitesse 4 km/h. Batterie 6 volts livrée avec rechargeur sur secteur. Démarrage au pied.
86 x 41 x 51 cm. . . . . F 900,00

□**PT1512-010**   **"1855 White"**   **Ertl**   **39.5" long**
Last of the Oliver pedal tractors. This model reflects the merger of the Oliver Corporation into the White Company as can be seen by the "White" name on the hood decal.

□**PT1850-004 "Renault Turbo"**   **Ampafrance 86 cm long**
Same pattern as -001. Yellow in place of Renault orange. Six volt battery powered.

33. Tracteur Renault à chaîne.
cm . . . . F 290,00

□**PT1805-002**   **"Turbo TX 133-14"**   **Falk**   **29.5" long**
A neat plastic model of the late Renault tractor. This European version features orange pedals, cleated front tires and a dark belt over front of hood.

□**PT1805-005**   **"951-4"**   **Falk**   **36" long**
Plastic construction with fenders, vertical muffler and cleated front tires. Hood decal proclaims the "Tracto-control" feature of the real tractors.

285

☐**PT1805-006** **"751-4"** **Falk** **29" long**
Similar to -005 with model designation change.

Umbrellas with trademarks or logos of major tractor manufacturers were available for Eska made tractors. They include: AC; IH; Massey-Harris; John Deere; Oliver and Case;. Two variations of John Deere include the leaping deer with and without "Quality Farm Equipment" printed on canopy.

☐**PY2308-001** **"C-195"** **Roadmaster Corp.** **39.5" long**
Another model representing a lawn and garden type tractor, "Wheel Horse". This pressed steel and tubular steel construction with plastic hood, seat, and wheels is made by Roadmaster Corp. of Olney Ill. This model is not an exact promotional of the "Wheel Horse" but does resemble it fairly well. This same tractor can be found in retail stores as a "Roadmaster".

Those represented include:
(clockwise)
John Deere (100400-024) 1/16
ERTL
David Brown 30 TD Crawler
(040200-009) 1/16
Shackleton-Chad Valley
Fordson Major (DDN) (061719-035) 1/16 **Chad Valley**
Same DA 38 DT (190113-010)
1/12 ??
OTO C25C Crawler (152015-002)
1/20 **Machpi-Bologna**
OTO C25R (152015-001) 1/20
Machpi-Bologna
Guildner Toledo (072112-002)
1/20 **Cursor**
Deutz D40L (040521-025) 1/21
Cursor
Nuffield Universal (M-IV)
(142106-001) 1/16 **Denzil**
Skinner
Same 240DT (190113-009) 1/12 ??
David Brown 25D (040200-001)
1/16 **Denzil Skinnner**

By permission of FABIO ZUBINI of Italy

**SPECIAL MINIATURE TRACTORS WITH OPERATOR
MANUALS FOR THEIR REAL COUNTERPARTS**

The charts on the following pages are designed to aid collectors in identifying models based on these criteria:

Code number—This is an arbitrary number assigned to each model and/or variation. The first six numbers indicate the real tractor manufacturer:

Example 010300- 01=A   02=C   00=0
                 Allis Chalmers
        090803- 09=I   08=H   03=C
                 International Harvester Corporation

An easy reference chart follows that identifies the numbers, letters and manufacturer.

The next three numbers, sometimes followed by a letter indicate the model and/or variation.
-011A=Plastic   D-Series
-011B=Plastic Kit   D-Series

Model-This is either the actual represented or the model most nearly represented.

Example 010300-017A represents the B-110 Lawn & Garden Tractor.

If the number appears in parenthesis, this indicates the authors' interpretation of the model represented.

Example 010300-005 (WC) This is the probable model represented, although this number does not appear on the miniature.

Scale or size-A real tractor is a 1/1 scale, meaning that it is full size. If the tractor were 1/2 its real size, it would be a 1/2 scale. A popular scale in the United States is 1/16, meaning that the real tractor is sixteen times as large as the miniature.

In Europe, a popular scale is 1/32. A real tractor, then, would be 32 times the size of the miniature.

In some situations, where the scale is not known, the actual size, either in inches or centimeters, is given. There are a few, particularly the bottle decanters, where neither size nor scale is given, but the volume is used.

Type of material-This column gives the material from which which the miniature is made. The abbreviations are listed below:

CA   Cast aluminum
CI   Cast iron-a ferrous material with magnetic attraction
Cus Custom made
D   Diecast (Not a material but a manufacturing process)
G   Gloss
K   Korloy-a non-metallic material somewhat like cast iron
L   Lead
MS   Machined Steel
P   Plastic
PK   Plastic Kit
POR Porcelain
POT Pot metal
R   Rubber
SS   Stamped steel
T   Tin
W   Wood
WK   Wood Kit
Z   Zymac-a type of pot metal

Manufacturer-The manufacturer of the miniature,
The Toy Company

Stock number-The manufacturer's number assigned to identify a particular miniature.

Country-The country in which the miniature was manufactured or the country in which the toy company is headquartered.

Year-The exact or approximate year when the miniature was manufactured.

Remarks-Additional details—See glossary of terms.
WFE   Wide front end (axle)
Row crop   Narrow front

Individual Tractor Charts

CODE	MODEL	SCALE	MATER.	MANU.	STOCK #	CTRY.	YEAR	REMARKS
010300-000	**ALLIS-CHALMERS**							
☐ 010300-001A	HD-14	1/20	POT	??		USA	??	Crawler with side cable lifts for Baker blade; Paperweight.
☐ 010300-001B	HD-14	1/20	POT	??		USA	??	Crawler without blade; Paperweight.
☐ 010300-002A	(A)	1/12	CA or K	ROBERT GRAY	NONE	USA	71	WFE; Antique style standard model from the '20's and '30's. Separately cast driver.
☐ 010300-002B	(A)	1/12	K	ROBERT GRAY	NONE	USA	79	WFE; Same except 10M. Anniversary model.
☐ 010300-003	(U)	3"	CI	ARCADE		USA	34	WFE; Standard style; Cast in driver; Represents a 1929 model.
☐ 010300-004	(U)	5"	CI	ARCADE	2657-50	USA	34	WFE; Standard style; Cast in driver; Cast in "Allis-Chalmers" name.
☐ 010300-005	(WC)	6"	CI	ARCADE		USA	40	Row crop; Cast in driver; Cast in name; Represents a 1934 model.
☐ 010300-006	(WC)	7"	CI	ARCADE	3704	USA	40	Row crop, Separately cast, plated driver.
☐ 010300-007	(WC)	1/16	CI	DENT		USA	40	Row crop; Separately cast, painted driver. Cast in name.
☐ 010300-008	(WC)	1/16	P	AUBURN RUBBER		USA	50	Row crop; Molded in driver; Red and silver.
☐ 010300-009	(C)	1/12	D	AMERICAN PRECISION		USA	50	Row crop; "Goodyear" tires.
☐ 010300-010	(WD-45)	1/16	P	PRODUCT MINIATURE	KC-3	USA	52	Row crop; very good detail.
☐ 010300-011A	(D-SER)	1/25	P	STROMBECKER	D61-149	USA	60	WFE; Assembled; Grill variations; Also stock no. KC-1.
☐ 010300-011B	(D-SER)	1/25	PK	STROMBECKER	D61-149	USA	60	WFE; Kit; Grill variations; Also stock no. KC-1.
☐ 010300-012	HD-5	1/16	P	PRODUCT MINIATURE		USA	55	Crawler with Baker blade.
☐ 010300-013A	HD-14	1/20	POT			USA	??	Crawler with Baker blade; Overhead cable lift; Paperweight.
☐ 010300-013B	HD-14	1/20	POT			USA	??	Crawler without blade; Paperweight.
☐ 010300-014A	HD-14	1/20	POT			USA	??	Crawler with Baker blade; Side hydraulic lift arms; Paperweight.
☐ 010300-014B	HD-14	1/20	POT			USA	??	Crawler without blade; Paperweight.
☐ 010300-015	(WC)	6"	CA	ERTL		USA	45	Row crop; Cast in driver; Small exhaust stack on hood; Similar to 010300-005 by Arcade.
☐ 010300-016A	(D-SER. I)	1/16	D	ERTL	104	USA	60	WFE; Steerable; Orange with black grill bars and decal border; Headlights.
☐ 010300-016B	(D-SER. II)	1/16	D	ERTL	104	USA	61	WFE; Steerable; Beige wheels; Light grill trim; Headlights.
☐ 010300-016C	(D-SER. III)	1/16	D	ERTL	104	USA	64	WFE; Steerable; Long side decals; No headlights.
☐ 010300-016D	(D-SER. IV)	1/16	D	ERTL	104	USA	64	WFE; Same except without the air cleaner on hood.
☐ 010300-017A	B-110	1/16	D	ERTL	197	USA	67	Lawn and garden tractor.
☐ 010300-017B	B-112	1/16	D	ERTL	AC-197	USA	69	Lawn and garden tractor; With blade, mower and two-wheel trailer.
☐ 010300-018	LGT	1/16	D	ERTL	151	USA	72	Lawn and garden tractor; With blade and trailer.
☐ 010300-019A	190	1/16	D	ERTL	192	USA	65	WFE; Bar grill; Metal wheels; Either silver or black AC tracemark on grill decal.
☐ 010300-019B	190	1/16	D	ERTL	192	USA	66	WFE; Bar grill; Plastic wheels; Either silver or black AC trademark on grill decal.
☐ 010300-020	190XT	1/16	D	ERTL	192	USA	69	WFE; No bars on grill.
☐ 010300-021	190XT III	1/16	D	ERTL	AC-188	USA	71	WFE; Rops; Large flotation type tires front and rear; Called "Landhandler".
☐ 010300-022A	BIG ACE	1/16	D & P	ERTL	2703	USA	72	WFE; Super Rod pulling tractor; Large tires.
☐ 010300-022B	BIG ACE	1/16	D & P	ERTL	2703	USA	73	WFE; Super Rod pulling tractor; Small wheels; Black outline decals.
☐ 010300-023A	200	1/16	D	ERTL	AC-152	USA	72	WFE; Black trim; Air cleaner and exhaust stacks; Front weights.
☐ 010300-023B	200	1/16	D	ERTL	AC-152	USA	73	WFE; Same except no air cleaner.
☐ 010300-024A	7030	1/16	D	ERTL	1202	USA	74	WFE; Orange over maroon.
☐ 010300-024B	7040	1/16	D	ERTL	1201	USA	75	WFE; Orange over maroon.
☐ 010300-024C	7050	1/16	D	ERTL	1200	USA	73	WFE; Cab; Wide rear tires; Orange over maroon.
☐ 010300-024D	7060	1/16	D	ERTL	1200	USA	75	WFE; Cab; Wide rear tires; No air cleaner; Orange over maroon.
☐ 010300-025A	12-G	1/25	D	ERTL	AC-198	USA	67	Crawler with loader; Yellow.
☐ 010300-025B	FIAT-ALLIS 12-G-B	1/25	D	ERTL	198	USA	75	Crawler with loader; Yellow.
☐ 010300-026A	HD-15	4"	P	LIONEL		USA	??	Crawler with blade; A.C. orange with identifying decals.
☐ 010300-026B	HD-15	4"	P	LIONEL		USA	??	Crawler with blade; Yellow without identification.
☐ 010300-027	SCRAPER PAN	7-1/2"	P	LIONEL		USA	??	Earth mover.

	CODE	MODEL	SCALE	MATER.	MANU.	STOCK #	CTRY.	YEAR	REMARKS
☐	010300-028	CRAWLER	1/80	D	MERCURY	517	I	61	Crawler with blade.
☐	010300-029	260 SCRAPER PAN	1/74	D	LESNEY	K-6	GB	61	Earth mover; Articulated.
☐	010300-030	(D-SERIES)	12"	P	EMPIRE		HK	??	Row crop; Molded in driver.
☐	010300-031	7045	1/64	D	ERTL	1623	USA	78	WFE; Cab.
☐	010300-032A	7045	1/16	D	ERTL	1201	USA	78	WFE; Cab; No air cleaner; Orange over black. Also available without cab.
☐	010300-032B	7060	1/16	D	ERTL	1208	USA	78	WFE; Cab; No air cleaner; Orange over black.
☐	010300-032C	7080	1/16	D	ERTL	1218	USA	79	WFE; Cab; No air cleaner; Duals; Orange over black.
☐	010300-032D	7080	1/16	D	ERTL	218	USA	79	WFE; Duals; Toy Farmer National Farm Toy Show Tractor 11/9/79.
☐	010300-033	(WC)	1/12	CA	SCALE MODELS	No. 1	USA	78	Row crop; "Steel" wheels; No. 1 in the JLE Collector Series; Represents a 1934 model; 3000 made.
☐	010300-034A	B	1/32	WM KIT	BRIAN PARKS	T-1	GB	81	WFE; Represents a 1939 British model; Adjustable front axle.
☐	010300-034B	B	1/32	WM KIT	BRIAN PARKS	T-2	GB	81	WFE; Represents a 1930 American model; Fixed, non-adjustable front axle.
☐	010300-035	WC	1/16	CUS	DENNIS PARKER	NONE	USA	80	Row crop; Represents a 1938 model; Very good detail.
☐	010300-036A	(WC)	1/16	CA	EARL JERGENSEN	NONE	USA	80	Row crop; No driver; Either on "steel" or rubber. Represents a '30's model: Limited production.
☐	010300-036B	(WC)	1/16	CA	EARL JERGENSEN	NONE	USA	80	Row crop; separate driver; Represents a '30's model; Limited production: Either on "steel" or rubber.
☐	010300-037A	(A)	1/12	CA	ROBERT GRAY	NONE	USA	71	WFE; Standard style tractor: "Steel" wheels.
☐	010300-037B	(A)	1/12	CA	ROBERT GRAY	NONE	USA	79	WFE; Standard style tractor; Separately cast gold colored drive: Rubber; Part of Robert Gray's 10th Anniversary set.
☐	010300-038	7045	1/32	PK	IMAI	B-971	J	81	WFE; Represents and off-road pulling tractor.
☐	010300-039	8550	1/32	D	ERTL	1213	USA HK	81	WFE; Four-wheel-drive; Articulated; Cab; Duals.
☐	010300-040	FORKLIFT	1/36	D	CORGI	409	GB-HK	81	Industrial fork lift with pallets.
☐	010300-041	8070	1/64	D	ERTL	1819	USA	82	WFE; Cab; Some very slight cab casting variations exist.
☐	010300-042A	U	1/16	CUS	MARBIL	NONE	GB	82	WFE; Represents a '30's tractor; "Steel" wheels.
☐	010300-042B	U	1/16	CUS	MARBIL	NONE	GB	83	WFE; Represents a '30's model; Rubber tires; Limited production.
☐	010300-043A	8030	1/16	D	ERTL	1220	USA	82	WFE; Front-wheel-assist; Cab; Collectors Series I; Reno, Feb. 82.
☐	010300-043B	8010	1/16	D	ERTL	1221	USA	82	WFE; Front-wheel-assist; Cab; Collectors Series I; Reno, Feb. 82.
☐	010300-043C	8030	1/16	D	ERTL	1220	USA	83	WFE; Cab; Dual wheels.
☐	010300-043D	8010	1/16	D	ERTL	1221	USA	83	WFE; Front-wheel-assist; Cab.
☐	010300-044A	4W-305	1/32	D & P	ERTL	1225DO	USA	83	WFE; Four-wheel-drive; Cab; Articulated.
☐	010300-044B	4W-305	1/32	D & P	ERTL	1225DA	USA	83	WFE; Four-wheel-drive; Cab; Duals Collectors Series.
☐	010300-045	HD-20	1/25	P	TRIANG MINIC		GB	56	Crawler with blade; Orange.
☐	010300-046	BIG ORANGE	750 ML	PROC	PACESETTER	NO. 4	USA	84	WFE; Cab; Porcelain and plastic decanter; 6000 made.
☐	010300-047	BIG ORANGE	50 ML	PROC	PACESETTER	NO. 4	USA	84	WFE; Cab; Porcelain and plastic decanter; 2000 made.
☐	010300-048	RC	1/16	CA	SCALE MODELS	NO. 12	USA	84	Row crop; "Steel" wheels; No. 12 in the JLE Ertl Collector Series; 5000 made.
☐	010300-049	A	1/12	WOOD	GUBBELS	NONE	USA	80	WFE; Rubber.
☐	010300-050	UC	1/12	WOOD	GUBBELS	NONE	USA	80	Row crop; Rubber.
☐	010300-051	U	1/12	WOOD	GUBBELS	NONE	USA	80	WFE; Rubber.
☐	010300-052	RC	1/12	WOOD	GUBBELS	NONE	USA	80	Row crop; Rubber.
☐	010300-053	W	1/12	WOOD	GUBBELS	NONE	USA	80	Row crop; "Steel".
☐	010300-054	WF UNSTYLED	1/12	WOOD	GUBBELS	NONE	USA	80	WFE; Rubber.
☐	010300-055	WF STYLED	1/12	WOOD	GUBBELS	NONE	USA	80	WFE; Rubber.
☐	010300-056	WC STYLED	1/12	WOOD	GUBBELS	NONE	USA	80	Row crop; Rubber.
☐	010300-057	WC UNSTYLED	1/12	WOOD	GUBBELS	NONE	USA	80	Row crop; Rubber.
☐	010300-058	B	1/10	WOOD	GUBBELS	NONE	USA	80	Row crop; Rubber.
☐	010300-059	C	1/12	WOOD	GUBBELS	NONE	USA	80	Row crop; Rubber.
☐	010300-060	CA	1/12	WOOD	GUBBELS	NONE	USA	80	WFE; Rubber.
☐	010300-061	D-10	1/12	WOOD	GUBBELS	NONE	USA	80	WFE; Rubber; Available in Series I, II and III.
☐	010300-062	D-12	1/12	WOOD	GUBBELS	NONE	USA	80	WFE; Rubber; Available in Series I, II and III.
☐	010300-063	WD	1/12	WOOD	GUBBELS	NONE	USA	80	Row crop; Rubber.
☐	010300-064	WD-45 LPG	1/12	WOOD	GUBBELS	NONE	USA	80	WFE; LPG fuel tank.
☐	010300-065	WD-45 DIESEL	1/12	WOOD	GUBBELS	NONE	USA	80	WFE.
☐	010300-066	WD-45 GAS	1/12	WOOD	GUBBELS	NONE	USA	80	WFE.

CODE	MODEL	SCALE	MATER.	MANU.	STOCK #	CTRY.	YEAR	REMARKS
☐ 010300-067	D-19 PROPANE	1/12	WOOD	GUBBELS	NONE	USA	80	WFE; LPG fuel tank.
☐ 010300-068	D-19 DIESEL	1/12	WOOD	GUBBELS	NONE	USA	80	WFE.
☐ 010300-069	D-19 GAS	1/12	WOOD	GUBBELS	NONE	USA	80	WFE.
☐ 010300-070	D-21	1/12	WOOD	GUBBELS	NONE	USA	80	WFE.
☐ 010300-071	D-21 SERIES II	1/12	WOOD	GUBBELS	NONE	USA	80	WFE.
☐ 010300-072	210	1/12	WOOD	GUBBELS	NONE	USA	80	WFE.
☐ 010300-073	220	1/12	WOOD	GUBBELS	NONE	USA	80	WFE.
☐ 010300-074	25-40	1/10	WOOD	KRUSE	NONE	USA	83	WFE; "Steel" wheels.
☐ 010300-075	WC	1/10	WOOD	KRUSE	NONE	USA	81	Row crop; Rubber.
☐ 010300-076	6-12	1/10	WOOD	KRUSE	NONE	USA	80	WFE; "Steel" wheels.
☐ 010300-077	18-30	1/10	WOOD	KRUSE	NONE	USA	83	WFE; "Steel" wheels.
☐ 010300-078	20-35	1/10	WOOD	KRUSE	NONE	USA	80	WFE; "Steel" wheels.
☐ 010300-079	A	1/10	WOOD	KRUSE	NONE	USA	80	WFE; Rubber.
☐ 010300-080	WC	1/10	WOOD	KRUSE	NONE	USA	79	Row crop; "Steel" wheels.
☐ 010300-081	B	1/12	CA	A.T. & T. COLL	NONE	USA	82	WFE.
☐ 010300-082	UC	1/16	CA	A T & T COLLECTABLES	NONE	USA	83	WFE; "Steel" wheels.
☐ 010300-083	440	1/16	CUS	GARY ANDERSON	NONE	USA	83	WFE; Four-wheel-drive; Articulated; Cab; Duals.
☐ 010300-084	??	1/87	P	??			??	Crawler; Crude.
☐ 010300-085	??	1/87	P	??			??	Scraper pan; Crude; Also used in a glass ball paperweight.

010718-000 AGRALE

☐ 010718-001	4300	1/25	D & P	ARPRA LTD	54	BR	83	WFE; Front-wheel-assist; A small tractor.

010800-000 ARTHUR HAMMER

☐ 010800-001	TRACTOR		P	ARTHUR HAMMER		D	??	WFE; Available with a two-wheel trailer.

011212-000 ALLCHIN

☐ 011212-001	7-32	1/80	D	LESNEY	Y-1	GB	55	WFE; Represents a 1925 steam traction engine.
☐ 011212-002	7-32	1/76	D KIT	ABS-MODELS	R-00 (9)	GB	76	WFE; Similar to 011212-001; Canopy optional.
☐ 011212-003	ROYAL CHESTER	1/32	WM KIT	WILLIS FINECAST		GB	??	WFE; Steam traction engine. Variations include agricultural engine, road engine and showman's engine.
☐ 011212-004	ROYAL CHESTER	1/160	WM	DG MODELS		GB	??	WFE; Steam traction engine.
☐ 011212-005	ALLCHIN	2 mm		ROWLAND PRODUCTS		GB	??	WFE; Steam traction engine.

011600-000 AGRIPOWER   MENNARD   THESE CASTINGS NOT EXACT REPRESENTATIONS OF THE REAL TRACTOR.

☐ 011600-001A	7000	1/16	D	SCALE MODELS		USA	82	WFE; Represents the Argentine Fiat; Tan and Brown.
☐ 011600-001B	8000	1/16	D	SCALE MODELS		USA	82	WFE; Same except decals and color; Tan and blue.
☐ 011600-001C	9000	1/16	D	SCALE MODELS		USA	83	WFE; Same except decals and color; Tan and brown.
☐ 011600-001D	11000	1/16	D	SCALE MODELS		USA	83	WFE; Same except decals and color; Tan and blue.

011818-000 ARROW

☐ 011818-001	AGRICULTURAL	1/12	CUS	KARSLAKE	NONE	GB	81	WFE; "Steel" wheels; This is a model maker's composite creation based upon both old and new concepts in agricultural tractors.

012119-000 AUSTIN

☐ 012119-001	AUSTIN	1/16	CUS	MARBIL	NONE	GB	80	WFE; "Steel" wheels; Very limited production.

012200-000 AVERLING & PORTER

☐ 012200-001	STEAM TRACTION ENGINE	1/43	WM KIT	PHANTOM MODELS	A-19	F	84	WFE; Represents an antique steam traction engine.

012205-000 AVERY

☐ 012205-001	??	4-1/2"	CI	HUBLEY		USA	20	WFE; Rounded radiator; "Avery" name cast on sides.
☐ 012205-002	??	4-1/2"	CI	ARCADE		USA	29	WFE; Square radiator; No name.
☐ 012205-003	??	4-1/2"	CI	BRUBAKER, PETERSON, WHITE, IRVIN	NONE	USA	64	WFE; Reproduction of Hubley Avery 012200-001. Irvin's model has name cast in it.
☐ 012205-004	UNDERMOUNT	1/25	CA	IRVIN	NONE	USA	78	WFE; Steam traction engine; Called "Undermount" because engine is located under the steam boiler.
☐ 012205-005	(A)	1/16	LEAD	TOM HOFFMAN	NONE	USA	80	Row crop; Single front wheel; Limited production: 110 made.

012220-000 AVTO

☐ 012220-001	(MTZ)	1/16	P	MINILUXE ?		F	75	WFE; Front-wheel-assist; Cab; Cream and red; USSR export model of Belarus; See also Belarus.

020111-000 BAKER

☐ 020111-001A	21-75 SPECIAL	1/25	CA	IRVIN	NONE	USA	83	WFE; Steam traction engine; Blue trim.
☐ 020111-001B	21-75	1/25	CA	IRVIN	NONE	USA	83	WFE; Steam traction engine; Red trim.

020120-000 BATES

☐ 020120-001	40 STEEL MULE	1/16	CI	VINDEX	4	USA	30	Crawler; Solid cast wheels and undercarriage; Separately cast plated driver.

CODE	MODEL	SCALE	MATER.	MANU.	STOCK #	CTRY.	YEAR	REMARKS
020121-000	**BAUTZ**							
☐ 020121-001	240	1/32	D				??	WFE; Promo.
☐ 020121-002	300	1/30	P	CURSOR		D	??	WFE; Promo: Seat back rails on both rear fenders.
020401-000	**BEAUCE-FLANDRE**							
☐ 020401-001	WHEEL TRACTOR	1/32	D	SOLIDO	85	F	50	WFE; Clockwork; Has "Solido" name cast on sides.
☐ 020401-002	CRAWLER TRACTOR	1/32	D	SOLIDO	87	FR	50	Crawler; Clockwork; Has "Solido" name cast on sides.
020412-000	**BELARUS**							
☐ 020412-001	420 (MTZ 52)	1/16	P	MINILUXE ?		F	75	WFE; Front-wheel-assist; Cab; USSR model; See also Avto.
☐ 020412-002	(MTZ-62)	1/43	D	YAXON		USSR	81	WFE; Front-wheel-assist; Red, blue or green; With or without "Belarus" inscription on sides; The USSR utilizes other countries in the production of selected products, however, stipulated that the USSR be shown as the location of manufacture; The cab of this miniature has "Made in USSR" embossed on it.
020907-000	**BIG BUD**							
☐ 020907-001	360/30	1/16	CA	TRUMM-DEBAILLIE	1 of 1100	USA	84	WFE; Four-wheel-drive; Duals; Articulated; Tilting cab and hood; 1100 made. "Limited Edition - 1 of 1100"
021200-000	**BLACKHAWK**							
☐ 021200-001	40	1/16	CA	BEN SIEGEL	NONE	USA	83	WFE or row crop.
021201-000	**BLAW KNOX**			**THESE TRACTORS ARE OLIVER CLETRACS**				
☐ 021201-001	CRAWLER	1/43	D	DINKY	561	GB	48	Crawler with closed engine; Variations: No. 961 with blade. No. 963 without blade; No. 885 with blade (made in France).
☐ 021201-002	CRAWLER	1/43	P	DINKY	961	GB	64	Crawler with open engine; Blade.
☐ 021201-003	CRAWLER	1/43	D	MOKO-LESNEY		GB	51	Crawler; With cast-in driver.
☐ 021201-004	CRAWLER	1/50	D	SALCO		GB	??	Crawler with blade.
☐ 021201-005	CRAWLER	1/50	D	JOAL	210	E	??	Crawler with blade.
☐ 021201-006	CRAWLER	1/43	D	DINKY		GB	48	Crawler; Early style.
021322-000	**B M VOLVO**							
☐ 021322-001	800	1/66	D	HUSKY	34	GB	67	WFE.
☐ 021322-002	800	1/66	D	CORGI JR		GB-HK	??	WFE; Very similar to 021322-001 by Husky.
☐ 021322-003	800	4-3/4"	P	TOMITE			75	WFE; Cab.
☐ 021322-004	T 650	1/50	D	Y-DIAPET	6666-151	J	79	WFE; Cab.
☐ 021322-005A	BM 2654	1/32	D & P	BRITAINS LTD	9521	GB	80	WFE; Front-wheel-assist; Cab.
☐ 021322-005B	BM 2654	1/32	D & P	BRITAINS LTD	9590	GB	83	WFE; Yellow; Industrial version with snow plow; Cab; Autoway Series No. 9880.
☐ 021322-006	BM	1/43	D & P	YAXON	033	I	83	WFE; Front-wheel-assist; Half cab.
☐ 021322-006	BM	1/32	D & P	YAXON	110	I	84	WFE; Front-wheel-assist; Full cab.
☐ 021322-007	BM	1/32	D & P	YAXON	110	I	84	WFE; Front-wheel-assist; Full cab.
☐ 021322-008	BM	13 cm		??			??	WFE.
021512-000	**BOLINDER MUNKTELL**							
☐ 021512-001	??	20 cm	P	SUEDE PENDING		D	??	WFE.
021600-000	**BUFFALO-PITTS**							
☐ 021600-001	??	1/25	KOR	ROBERT GRAY	NONE	USA	74	WFE; Antique steam traction engine; Available with thresher and water wagon.
022118-000	**BURRELL**							
☐ 022118-001	??	1/8	METAL	MAXWELL HEMMENS		GB	??	WFE; Showman's steam traction engine; Excellent detail; Very limited production.
022211-000	**BUKH**							
☐ 022211-001	D-30 DIESEL	1/43	D	CHICO TOYS	17	CO	76	WFE; Copy of Villmer Bukh 022211-002.
☐ 022211-002	D-30	1/43	D	VILMER		DK	60	WFE.
030118-000	**CARRARO**							
☐ 030118-001A	88.4	1/24	D & P	BARLUX	73049	I	80	WFE; Front-wheel-assist; ROPS.
☐ 030118-001B	88.4	1/24	D & P	SCAME-GIODI	73049	I	81	WFE; Similar to 030118-001A; New name for manufacturer.
☐ 030118-001C	JUMBO (88.4)	1/24	D & P	SCAME GIODI	73030	I	82	WFE; Cab; Yellow.
☐ 030118-001D	920	1/24	D & P	SCAME GIODI	73049	I	84	WFE; Front-wheel-assist; Similar to 030118-001A.
☐ 030118-001E	BULL 34 (88.4)	1/24	D & P	SCAME GIODI	73034	I	84	WFE; Front-wheel-assist; With side dump wagon.
☐ 030118-001F	MASTER 33 (88.4)	1/24	D & P	SCAME GIODI	73033	I	84	WFE; Front-wheel-assist; Cab; Blue; With high side wagon.
☐ 030118-002	6500	1/20	P	AGFA DI FAVERO	200	I	70	WFE; Promo.
030119-000	**CASE**							
☐ 030119-001	(L)	1/16	CI	VINDEX	36	USA	30	WFE; 1929 Standard style tractor; Separately cast plated driver.
☐ 030119-002A	(L)	1/16	CA	OLD TIME TOYS	NONE	USA	68	WFE; Reproduction of the Vindex Case (L) 030119-001.

CODE	MODEL	SCALE	MATER.	MANU.	STOCK #	CTRY.	YEAR	REMARKS
☐ 030119-002B (L)		1/16	CA	PIONEER TRACTOR WORKS		USA	80	WFE; Reproduction of Vindex Case (L) 030119-001.
☐ 030119-003A (SC)		1/16	P	MONARCH PLASTIC		USA	50	Row crop; Orange; The real Case SC was introduced in 1939.
☐ 030119-003B (SC)		1/16	P	MONARCH PLASTIC		USA	51	Row crop; Same except having fenders.
☐ 030119-004A CASOMATIC		1/16	P	JOHAN		USA	56	Row crop; Beige over orange.
☐ 030119-004B	800 CASOMATIC	1/16	P	JOHAN		P	57	Row crop; Same except having "800" number.
☐ 030119-005A	930 CK	1/16	D	ERTL	204	USA	63	WFE; "Wheatland" style with fenders over the rear wheels; Beige over orange; front wheel variations include wheels with or without slots.
☐ 030119-005B	1030	1/16	D	ERTL	204	USA	63	WFE; Same except with 1030 decals.
☐ 030119-006	1030 CK	1/16	D	ERTL	204	USA	67	WFE; Same except having flat fenders and front weights.
☐ 030119-007A	1070 AK	1/16	D	ERTL	200	USA	69	WFE; Desert sunset over orange.
☐ 030119-007B	1070 AK	1/16	D	ERTL	210	USA	69	WFE; Cab; Dual rear wheels.
☐ 030119-007C	1070 AK DEMO.	1/16	D	ERTL	210	USA	70	WFE; Cab; Black, orange and gold; Called the "GOLDEN HARVESTER DEMONSTRATOR" Very limited production.
☐ 030119-007D	1070 AK 451 DEMO.	1/16	D	ERTL	210	USA	71	WFE; Cab; Black with gold trim; Called the "BLACK KNIGHT DEMONSTRATOR".
☐ 030119-008A	1270 AK 451	1/12	D	ERTL	215	USA	72	WFE; Larger castings.
☐ 030119-008B	1270 AK 451	1/16	D	ERTL	216	USA	72	WFE; Cab.
☐ 030119-008C	1370 AK 504 TURBO	1/16	D	ERTL	216	USA	72	WFE; Cab; Decal variations including large or small 1370 number.
☐ 030119-008D	(1370) AGRI-KING	1/16	D	ERTL	216	USA	74	WFE; Color change to power red and white.
☐ 030119-008E	(1370) AGRI-KING	1/16	D	ERTL	262	USA	74	WFE; Cab.
☐ 030119-008F	SPIRIT OF 76	1/16	D	ERTL	217	USA	76	WFE; Cab; Red, white and blue with stars; Light blue.
☐ 030119-008G	SPIRIT OF 76	1/16	D	ERTL	217	USA	76	WFE; Same except dark blue.
☐ 030119-009A	??	1/25	CA	BRUBAKER, WHITE PETERSON, IRVIN	NONE	USA	64	WFE; Antique steam traction engine with canopy and driver; Also available with water wagon and thresher.
☐ 030119-009B	STEAM ROLLER	1/16	CA	IRVIN		USA	76	Roller front; A variation of the steam traction engine.
☐ 030119-009B	STEAM ROLLER	1/16	CA	IRVIN	NONE	USA	76	Roller front; A variation of the steam traction engine.
☐ 030119-010	CASE-D.B. 995	1/43	D	DINKY	305	GB	75	WFE; Cab; Red and white; Three-point-hitch.
☐ 030119-011	CASE D.B. 1412	1/25	D & P	NZG	156	D	76	WFE; ROPS; Red and white; Three-point-hitch.
☐ 030119-012	(1412)	1/25	D	NZG	159	D	76	WFE; Same as 030119-011 except without ROPS and front weights.
☐ 030119-013	(2670) TK	1/40	D	NZG	149-154	D	76	WFE; Four-wheel-drive; Four-wheel steering; Cab; Three-point-hitch.
☐ 030119-014	580C-CK	1/40	D	NZG		D	76	WFE; Industrial backhoe-loader; Case yellow and black.
☐ 030119-015	(2670) TK	1-1/4"	D IN L			USA	76	WFE; Paperweight lucite prism; 3-3/8" overall.
☐ 030119-016	(1000D)	1/82	D	LESNEY	16D	GB	69	Crawler with blade; Cab; Red and yellow.
☐ 030119-017	(1000D)	3-1/2"	D	LESNEY	K-17	GB	69	Crawler with blade; Cab; Red and yellow.
☐ 030119-018	580 CK	1/16	P	TOMY	20531	HK	67	WFE; Industrial backhoe-loader. Very good detail.
☐ 030119-019A	5808 CK	1/16	D	GESCHA	600	D	73	WFE; Industrial backhoe-loader; Yellow over orange engine and wheels.
☐ 030119-019B	5808 CK	1/16	D	GESCHA	600	D	75	WFE; Industrial backhoe-loader; All yellow.
☐ 030119-020	CRAWLER	1/16	P	TOMY ?		HK	65	Crawler/blade; Battery powered; Blade raises and lowers as crawler moves along. Rare.
☐ 030119-021A	CASE-DB 1412	1/32	D	CORGI	34	GB	77	WFE; Cab; Also available with tipping trailer.
☐ 030119-021B	CASE-D.B. 1412	1/32	D	CORGI	1112	GB	77	WFE; Cab; Has mounted J-F combine harvester.
☐ 030119-022	(CC)	1/16	CA	EARL JERGENSEN	NONE	USA	77	Row crop; Represents a 1929 model; "Steel" wheels; Very limited production.
☐ 030119-023	(AGRI-KING)	1/64	D	ERTL	1624	USA	78	WFE; Cab.
☐ 030119-024	859-B CRAWLER	1/35	D	NZG	176	D	78	Crawler/angle dozer blade; ROPS.

	CODE	MODEL	SCALE	MATER.	MANU.	STOCK #	CTRY.	YEAR	REMARKS
☐	030119-025A	2390	1/16	D	ERTL	268	USA	79	WFE; Small wheels; Cab.
☐	030119-025B	2590	1/16	D	ERTL	269	USA	79	WFE; Large wheels; Cab.
☐	030119-025C	2390 COLL. SER	1/16	D	ERTL	268	USA	79	WFE; Cab; Collectors Series; 1500 made.
☐	030119-025D	2590 COLL. SER	1/16	D	ERTL	269	USA	79	WFE; Cab; Large wheels; Collectors Series; 1500 made.
☐	030119-025E	2390 RECALL	1/16	D	ERTL	268	USA	79	WFE; Cab; These early production units were recalled because of defects.
☐	030119-025F	2590 RECALL	1/16	D	ERTL	269	USA	79	WFE; Cab; These early production units were recalled because of defects.
☐	030119-025G	2590	1/16	D	ERTL		USA	81	WFE; Front-wheel-assist; Cab; Toy Farmer National Farm Toy Show Tractor. 1000 made.
☐	030119-026	12-20 CROSSMOUNT	1/16	CUS	KEN CONKLIN	1	USA	81	WFE; Represents a 1929 model; 150 made.
☐	030119-027	(AGRI-KING)	1/64	P	?		?	??	WFE; Similar to Ertl 030119-027 (Agri-King).
☐	030119-028	580C CK	1/35	D	CONRAD		D	76	WFE; Industrial backhoe-loader.
☐	030119-029A	580D CK	1/35	D	CONRAD	2331	D	81	WFE; Industrial backhoe-loader.
☐	030119-029B	580D CK SILVER ANN.	1/35	D	CONRAD		D	81	WFE; Industrial backhoe-loader; Silver and black.
☐	030119-030	580F CK	1/35	D	CONRAD	293	D	81	WFE; Industrial backhoe-loader.
☐	030119-031	CASE-DROTT 50 EXC.	1/35	D	NZG		D	81	Industrial excavator.
☑	030119-032	980 B EXCAVATOR	1/35 ?	D	NZG		D	81	Industrial excavator.
☐	030119-033	1845 UNI-LOADER	1/35	D	NZG	196	D	80	Skid-Steer loader.
☐	030119-034A	4890	1/32	D	ERTL	A691	USA	82	WFE; Four-wheel-drive; Cab; Collectors Series.
☐	030119-034B	4890	1/32	D	ERTL	1691	USA	82	WFE; Four-wheel-drive; Cab.
☐	030119-035	980B	1/35	D	CONRAD	2961	D	80	Industrial excavator.
☐	030119-036	DROTT 50	1/35	D	CONRAD	2960	D	80	Industrial excavator.
☐	030119-037	740 WHEEL LOADER	1/35	D	??		S	81	Industrial wheel loader.
☐	030119-038A	2290	1/32	D	ERTL	1692	USA	82	WFE; Cab.
☐	030119-038B	2290 COLL. SER.	1/32	D	ERTL	1692	USA	82	WFE; Same except Collectors Series.
☐	030119-038C	2294	1/32	D	ERTL	1692TA	USA	83	WFE; Cab; Duals; Collectors inscription, "CASE 1983 LIMITED EDITION.
☐	030119-039A	1690	1/32	D	ERTL	1787	USA	82	WFE; ROPS.
☐	030119-039B	1690 COLL. SER.	1/32	D	ERTL	1787	USA	82	WFE; ROPS; Collector Series.
☐	030119-040A	CASE-D.B. 1690	1/32	D	ERTL	H787	USA	82	WFE; Cab.
☐	030119-040B	CASE-DB 1690 COL. SER.	1/32	D	ERTL	H787	USA	82	WFE; Cab; Collector Series.
☐	030119-040C	1690	1/32	D	ERTL	1717	USA	82	WFE; Cab.
☐	030119-041	2590	1/64	D	ERTL	1694	USA	82	WFE; Cab.
☐	030119-042	400	1/16	CUS	DENNIS PARKER	NONE	USA	82	Row crop; Represents a 1956 model; Very good detail.
☐	030119-043	VAC	1/16	CUS	PETE FREIHEIT	NONE	USA	82	Row crop; Excellent detail.
☐	030119-044	1930 (CC)	1/16	CA	SCALE MODELS	#8	USA	82	Row crop; "Steel" wheels; No. 8 in the JLE Collector Series. 3000 made.
☐	030119-045	(CASE) D.B. 2290	1/32	D	LONE STAR	1760	GB	82	WFE; Cab; Wheel variations include all black and red center.
☐	030119-046A	(DC)	1/16	CUS	LYLE DINGMAN	NONE	USA	81	Row crop; Excellent detail; Also a LPG model.
☐	030119-046B	(DC) LPG	1/16	CUS	LYLE DINGMAN	NONE	USA	84	Row crop; Excellent detail; LP Gas model with pressurized fuel tank.
☐	030119-047	D-STANDARD	1/16	CUS	LYLE DINGMAN	NONE	USA	84	WFE; Represents a 1942 model; Either on "steel" or rubber; Very good detail.
☐	030119-048	20-40	1/16	CA	SCALE MODELS	NO. 5	USA	83	WFE; Antique style steam traction engine; Threshers Series.
☐	030119-049	CROSSMOUNT 12-20	1/76	WM KIT	VM MINIATURES		GB	82	WFE; Antique style tractor; The model 12-20 is gray while the "A" model is green.
☐	030119-050A	300	1/16	CUS	BEN SIEGEL	NONE	USA	83	Row crop.
☐	030119-050B	300	1/16	CUS	BEN SEIGEL	NONE	USA	83	WFE; Beige over orange.
☐	030119-051A	2594 COLL. SER.	1/16	D	ERTL	266TA	USA	84	WFE; Cab; New color pattern that includes black; Collectors Series.
☐	030119-051B	2494	1/16	D	ERTL		USA	84	WFE; Cab; New color pattern that includes black
☐	030119-052A	3294 COLL. SER.	1/16	D	ERTL	266TA	USA	84	WFE; Front-wheel-assist; Cab; Collectors Series; New color pattern that includes black.
☐	030119-052B	3294	1/16	D	ERTL		USA	84	WFE; Front-wheel-assist; Cab; New color pattern that includes black.
☐	030119-053	2294	1/32	D	ERTL	261EO	USA	84	WFE; Front-wheel-assist; Cab.
☐	030119-054	(1594)	1/32	D	LONE STAR	1760	GB	84	WFE; Cab; White over black with red wheels and trim.
☐	030119-055	4890	1/35	D	CONRAD	5010	D	81	WFE; Four-wheel-drive; Cab; Three-point-hitch.
☐	030119-056	580F	1/40	C. KIT	PUBLI K		GB	??	WFE; Industrial backhoe-loader; Cardboard kit.
☐	030119-057	CC	1/10	WOOD	MARVIN KRUSE	NONE	USA	79	Row crop; "Steel" wheels.
☐	030119-058	CASOMATIC 800	1/16	CUS	GERRY WARNER	NONE	USA	84	Row crop; Steerable; Limited production. Also available as a kit.

	CODE	MODEL	SCALE	MATER.	MANU.	STOCK #	CTRY.	YEAR	REMARKS
☐	030119-059	SC	1/16	CUSTOM	BURKHOLDER-PTW	NONE	USA	78	Row crop; Very good detail; Serial numbered.
☐	030119-060A	4894	1/32	D	ERTL	262 TA	USA	84	Four-wheel-drive; Cab; Articulated; Collectors Series "FIRST EDITION"; 4000 made; New color pattern that includes black.
☐	030119-060B	4894	1/32	D	ERTL	262FO	USA	84	Four-wheel-drive; Cab; Articulated; New color pattern that includes black.
☐	030119-061	CC	1/12	CUSTOM	MARBIL	NONE	GB	80	Row crop; "Steel" wheels; Very limited production.
☐	030119-062	CASE (2594)	750 ML	PORC	PACESETTER	5	USA	84	WFE; Cab; Decanter; 6000 made.
☐	030119-063	CASE (2594)	50 ML	PORC	PACESETTER	5	USA	84	WFE; Cab; Decanter; 2400 made.
☐	030119-064	C	1/10	WOOD	KRUSE	NONE	USA	81	WFE; "Steel" wheels.
☐	030119-065	L	1/10	WOOD	KRUSE	NONE	USA	82	WFE; "Steel" wheels.
☐	030119-066	RC	1/16	CA	A.T. & T. COLL.	NONE	USA	82	Row crop; "Steel" wheels; Also with single front wheel.
☐	030119-067	LA	1/16	CUSTOM	LYLE DINGMAN	NONE	USA	84	WFE; Standard style; Excellent detail.
☐	030119-068A	15-45	1/16	CA	SCALE MODELS		USA	84	WFE; "Steel" wheels; Steam traction engine: "Midwest Threshers Reunion". 400 made.
☐	030119-068B	15-45	1/16	CA	SCALE MODELS		USA	84	WFE; Regular issue.
☐	030119-069	500	1/43	D	ERTL	2510	USA	85	WFE: Fenders over rear wheels; Represents a 1953 model.

030120-000 CASSANI

	CODE	MODEL	SCALE	MATER.	MANU.	STOCK #	CTRY.	YEAR	REMARKS
☐	030120-001	40 CV 1827	1/30				I	??	WFE; The first SAME tractor.

030801-000 CHASESIDE

	CODE	MODEL	SCALE	MATER.	MANU.	STOCK #	CTRY.	YEAR	REMARKS
☐	030801-001	??	1/87	P	ANGUPLAS	101E		??	WFE; Industrial tractor with front loader; Yellow.

031201-000 CLARK-BOBCAT

	CODE	MODEL	SCALE	MATER.	MANU.	STOCK #	CTRY.	YEAR	REMARKS
☐	031201-001A	M-700	1/24	D	GESCHA	401	D	75	Four-wheel skid steer loader; Black trim; ROPS.
☐	031201-001B	M-700	1/24	D	GESCHA	9420	D	77	Four-wheel skid steer loader; White trim; ROPS. Variations include one with no number designation and green stripe and another has the M-741 number with a blue stripe.
☐	031201-002	533 HYDROSTATIC	1/24	D	GAMA	9420	D	78	Four-wheel skid steer loader; Black trim; ROPS.
☐	031201-003	741 HYDROSTATIC	1/24	P & SS	TONKA	837087	USA	79	Four-wheel skid steer loader; Black trim; ROPS. One variation has only Hydrostatic without 741 number and green stripe while another variation has the 741 number with a blue stripe and the name "Tonka" on the wheels.

031204-000 CLETRAC — SEE ALSO OLIVER

	CODE	MODEL	SCALE	MATER.	MANU.	STOCK #	CTRY.	YEAR	REMARKS
☐	031204-001	CRAWLER	?	D	RONSON		USA	20	Crawler; Actually a cigarette lighter shaped like a Cletrac crawler.

031503-000 COCKSHUTT

	CODE	MODEL	SCALE	MATER.	MANU.	STOCK #	CTRY.	YEAR	REMARKS
☐	031503-001	30	1/16	D	ADVANCED PRODUCTS		USA	??	Row crop; Red and yellow; Non-steerable.
☐	031503-002	(540)	1/16	D	ADVANCED PRODUCTS		USA	??	WFE; Tan; Steerable; Three-point-hitch; Also, a tan over red version was made as a dealer award with a pen holder on the seat.
☐	031503-003	30	1/16	POT	LINCOLN TOYS		CAN	??	Row crop; Red and yellow.
☐	031503-004	30	1/16	D	LINCOLN TOYS		CDN	??	WFE; Red and yellow.
☐	031503-005A	1850	1/16	D	ERTL		USA	68	Row crop; Tan over red; Some had fenders; Canadian version of the Oliver 1850.
☐	031503-005B	1850	1/16	D	ERTL		USA	68	Row crop; Red with white wheels; Fenders.
☐	031503-006	30	1/16	P	KEMP PLASTIC CO.		USA	54	WFE; Red and yellow; Plastic with diecast front axle; Rare.
☐	031503-007	70	1/16	D	SCALE MODELS NUMBERED	SERIAL	USA	83	Row crop; Canadian version of the Oliver 70; "Steel" wheels; Represents a 1938 model. 5000 made. Serial numbered. Antique Series.
☐	031503-008	30	1/16	POT	??		USA	??	Row crop; Orange; Air cleaner and exhaust stacks.
☐	031503-009	30	1/16	CA	A.T. & T. COLL.	NONE	USA	83	Row crop; Also available with wide front end.
☐	031503-010	30	1/16	CA	A T & T COLL.	NONE	USA	83	Row crop; Also available with wide front end.

031512-000 COLORADO

	CODE	MODEL	SCALE	MATER.	MANU.	STOCK #	CTRY.	YEAR	REMARKS
☐	031512-001	CRAWLER	5''	D	QUIRALU		F	75	Crawler; Clockwork; Available also with blade or V-snowplow blade.

031514-000 CONTINENTAL

	CODE	MODEL	SCALE	MATER.	MANU.	STOCK #	CTRY.	YEAR	REMARKS
☐	031514-001	CD8	1/43	P	NOREV	111	F	60	Crawler with blade.
☐	031514-002	CD8	1/24	TIN & P	MONT BLANC	565	F	60	Crawler; Clockwork or battery.

031515-000 CO-OP

	CODE	MODEL	SCALE	MATER.	MANU.	STOCK #	CTRY.	YEAR	REMARKS
☐	031515-001	(E-3)	1/16	POT	ADVANCED PRODUCTS		USA	50	Row crop; Non-steerable; Orange; Predecessor to the Cockshutt line of tractors.
☐	031515-002	(E-3)	1/16	D	ADVANCED PRODUCTS		USA	51	Row crop; Steerable; Orange; Has exhaust stack, air cleaner stack and gear shift.

CODE	MODEL	SCALE	MATER.	MANU.	STOCK #	CTRY.	YEAR	REMARKS
☐ 031515-003A	E-3	1/16	CA	A.T. & T. COLL.	NONE	USA	83	Row crop.
☐ 031515-003B	E-3	1/16	CA	A.T. & T. COLL.	NONE	USA	83	WFE.
031521-000 COUNTY								
☐ 031521-001	FORDSON E27N CR.	1/16	CUS.	MARBIL	NONE	GB	83	Crawler; The County represents a Fordson conversion; Dark blue; Very limited production.
☐ 031521-002	SUPER 4-754	1/12	CUS.	DAVID SHARP	NONE	USA	84	WFE; Four-wheel-drive; Steerable: Three-point-hitch: Modified Ertl Ford 4000.
031522-000 COVENTRY-CLIMAX								
☐ 031522-001	FORK LIFT	1/35	D	DINKY	14c	GB	??	Fork lift.
032102-000 CUB CADET CORP.								
☐ 032102-000A	CCC (682)	1/16	D	ERTL	499	USA	82	Lawn and garden tractor; Yellow and white; Formerly owned by International Harvester Corporation.
☐ 032102-001B	CCC (682) COL. ED.	1/16	D	ERTL	499	USA	83	Lawn and garden tractor; Same except Collectors Version.
040106-000 DAF								
☐ 040106-001	PONY	1/43	D	LION CAR	45	NL	??	WFE; Transporter.
040109-000 DAIN						SEE ALSO JOHN DEERE		
☐ 040109-001	ALL WHEEL DRIVE	1/16	CUS.	FRANK HANSEN NUMBERED	Serial	USA	84	WFE; Single rear wheel: This miniature was made as a commemorative to the establishment that it, not the model "D" was the first John Deere tractor. Excellent detail.
040200-000 DAVID BROWN						SEE ALSO CASE		
☐ 040200-001	(25-D)	1/16	D	DENZIL SKINNER		GB	54	WFE; Steerable; Red; Excellent detail; Rare.
☐ 040200-002	990	1/43	D	DINKY	305	GB	64	WFE; Cab; Red and yellow; Three-point-hitch.
☐ 040200-003	990 SELECTOMATIC	1/43	D	DINKY	305	GB	66	WFE; Cab; Brown and white; Three-point-hitch.
☐ 040200-004	CROPMASTER	1/16	P	?		GB	50	WFE; Shields around driver's platform; Double seats; Rare; Promo.
☐ 040200-005	CROPMASTER	1-1/2"	P	MERIT		GB	50	WFE; Six part plastic puzzle; A premium in a corn flakes cereal box during the '50's in England.
☐ 040200-006	VAK CROPMASTER	1/72	PK	AIRFIX	06002-4	GB	75	WFE; Aircraft tow tractor; Part of a model airplane kit.
☐ 040200-007	1690	1/32	D	ERTL	1717	USA	82	WFE; Same as Case 1690.
☐ 040200-008	(1690)	1/32	D	LONE STAR	1760	GB	83	WFE; White with red or black wheels.
☐ 040200-009	30 TD CRAWLER	1/16	D	SHACKLTON-CHAD VALLEY		GB	56	Crawler; Red; Clockwork with fore-ward-reverse; Rare.
☐ 040200-010	??	1/70	P	?	GB?	??		Red.
040521-000 DEUTZ								
☐ 040521-001	D6006	1/32	D	MINI-AUTO	297	D	65	WFE; See also Ziss R. W. Modelle.
☐ 040521-002	60PS	1/90	D	SCHUCO	752	D	62	WFE; Red or green.
☐ 040521-003A	??	1/90	D	SCHUCO	753	D	62	Crawler; Red or yellow.
☐ 040521-003B	??	1/90	D	SCHUCO	754	D	62	Crawler with blade; Red or yellow.
☐ 040521-004	DM-55	1/25	P	TROL	5113	BR	64	WFE; Green or orange; Rare.
☐ 040521-005A	06 SERIES	1/32	D	ZISS R.W. MODELLE		D	70	WFE; Dark green and gray with orange seat.
☐ 040521-005B	06 SERIES	1/32	D	ZISS R.W.		D	75	WFE; Light green and brown with red seat.
☐ 040521-006	??	1/20	P	ARTHUR HAMMER	2452	D	??	WFE; See also 040521-019 by Hausser.
☐ 040521-007	?	1/25	P	CURSOR		D	??	WFE.
☐ 040521-008A	D 100 06A	1/29	D	GAMA	424	D	76	WFE; Front-wheel-assist.
☐ 040521-008B	D 100 06A	1/29	D	GAMA	426	D	78	WFE; Front-wheel-assist; ROPS.
☐ 040521-009A	INTRAC 2005	1/28	D	GAMA	420	D	76	WFE; Front-wheel-assist; Front and rear three-point-hitch; Rear dumpbed.
☐ 040521-009B	INTRAC 2005	1/28	D	GAMA	4225	D	76	WFE; Same except with dump trailer.
☐ 040521-009C	INTRAC 2005	1/28	D	GAMA	2308	D	80	WFE; Same except with sprayer tank.
☐ 040521-009E	INDUSTRIAL INTRAC	1/28	D	GAMA	425	D	80	WFE; Orange; (4225 with brush).
☐ 040521-009F	INTRAC 2005	1/28	D	GAMA		D	??	WFE; Front-wheel-assist; Same as 040521-009C except with sprayer booms.
☐ 040521-010	?	1/90	P	WIKING	T-6	D	??	WFE; With driver; Good detail; red or gray.
☐ 040521-011A	?	1/90	P	WIKING	383	D	64	WFE; With driver; Good detail; Green.
☐ 040521-011B	?/ROPS	1/90	P	WIKING	383	D	81	WFE; ROPS.
☐ 040521-012A	GAMA-DEUTZ	1/19	D	GAMA	432	D	77	WFE; Front-wheel-assist; Front weights; No cab.
☐ 040521-012B	(DEUTZ)	1/19	D	GAMA	4321	D	77	WFE; Same except with front loader.
☐ 040521-013A	DX-110	1/32	D & P	BRITAINS LTD	9526	GB	78	WFE; Front-wheel-assist; Cab; Rear hitch lever inside the cab.
☐ 040521-013B	DX-110	1/32	D & P	BRITAINS LTD	9526	GB	80	WFE; Front-wheel-assist; Cab; Rear hitch lever outside the cab.
☐ 040521-013C	DEUTZ-FAHR DX-110	1/32	D & P	BRITAINS LTD	9526	GB	84	WFE; Same except different decals.
☐ 040521-013D	DEUTZ-FAHR DX-110	1/32	D & P	BRITAINS LTD	9530	GB	84	WFE; Same except orange dual wheels both front and rear.

	CODE	MODEL	SCALE	MATER.	MANU.	STOCK #	CTRY.YEAR	REMARKS
☐	040521-013E	DEUTZ-FAHR DX-82	1/32	D & P	BRITAINS LTD	9530	GB 84	WFE; Front-wheel-assist; Cab; Same except decals.
☐	040521-013F	DX-110 INDUSTRIAL	1/32	D & P	BRITAINS LTD	9880	GB 84	WFE; Same except with dump trailer and industrial yellow; Autoway Series.
☐	040521-014A	D6206	1/32	D	HAUSSER	4425	D 78	WFE; Rear lift; Light green; ROPS.
☐	040521-014B	D6206	1/32	D	HAUSSER	4428	D 78	WFE; Same except no ROPS.
☐	040521-015	(D-15)	1/32	P	HAUSSER ELASTOLIN	4418	D 68	WFE; Green; 1950 model.
☐	040521-016	INTRAC 2003	1/32	D	HAUSSER	4480	D 80	WFE; Front and rear hitches.
☐	040521-017	06 SERIES	1/87	P	BRUDER		D 74	WFE; Red and yellow or red and green.
☐	040521-018	06 SERIES	4-1/2"	D	GAMA		D ??	WFE; Orange with a green cab.
☐	040521-019	FIL 514	1/20	P	HAUSSER	2452	D ??	WFE.
☐	040521-020	D10006		D	GAMA	2308	D ??	WFE; ROPS; With or without mower (side mounted).
☐	040521-021A	06 SERIES	8"	P	BRUDER		D 81	WFE; No cab.
☐	040521-021B	06 SERIES	8"	P	BRUDER		D 81	WFE; Cab.
☐	040521-021C	06 SERIES	8"	P	BRUDER		D 81	Cab; Front blade.
☐	040521-022	D5006	14-1/2"	P	?		78	WFE; With two-wheel trailer.
☐	040521-023	DX 230 POWER-MATIC	1/12	P	GAMA	3305	D 82	WFE; ROPS.
☐	040521-024A	DEUTZ FAHR DX 86	1/32	D	SIKU	2850	D 83	WFE; Front-wheel-assist; Rear lift; Cab.
☐	040521-024B	DEUTZ FAHR DX 4-70	1/32	D	SIKU	2950 ?	D 84	WFE; Front-wheel-assist; Rear lift; Cab; Duals.
☐	040521-025	D-40L	1/21 ?	P	CURSOR		D 60	WFE; Promo.

040920-000 DITCH WITCH

	CODE	MODEL	SCALE	MATER.	MANU.	STOCK #	CTRY.YEAR	REMARKS
☐	040920-001A	4010	1/40	D	ERTL	1384	USA 84	Industrial trencher with blade and backhoe.
☐	040920-001B	4010	1/40	D	ERTL	1484	USA 84	Industrial trencher with blade.

041505-000 DOE TRIPLE-D

	CODE	MODEL	SCALE	MATER.	MANU.	STOCK #	CTRY.YEAR	REMARKS
☐	041505-001	TRIPLE-D	1/32	WM KIT	BRIAN PARKS	T13	GB 84	WFE; Two Fordson Majors connected in tandem.
☐	041505-002	TRIPLE-D	1/16	D	CHAD VALLEY ?		GB ??	WFE; Two Fordson Majors connected in tandem; Conversion thought to be done by someone other than Chad Valley; Rare; May be one of a kind.

041710-000 DONG FONG

	CODE	MODEL	SCALE	MATER.	MANU.	STOCK #	CTRY.YEAR	REMARKS
☐	041710-001	2-WHEEL TRACTOR	6"	TIN	MS	857	CH 80	Garden type tractor; Clockwork; Tin driver with plastic head.

042120-000 DUTRA

	CODE	MODEL	SCALE	MATER.	MANU.	STOCK #	CTRY.YEAR	REMARKS
☐	042120-001	D-4K-B	9"	TIN	??		HU 65	WFE; Four-wheel-drive; Clockwork; Stamped steel. Rear drawbar serves as windup crank.
☐	042120-002	D4K	1/87	P	ESPEWE PLASTI-CART	1090	EG 79	WFE; Four-wheel-drive; Red, blue or green.
☐	042120-003	D4K	1/16 ?	P	??	1805L	HU 80	WFE; Four-wheel-drive; "DEBRECENI MUANYAG IPAR SZOVETKEZET".

050107-000 EAGLE

	CODE	MODEL	SCALE	MATER.	MANU.	STOCK #	CTRY.YEAR	REMARKS
☐	050107-001	GASOLINE ENGINE	1/10	CA	IRVIN	NONE	USA 77	Represents an antique gasoline engine.

050118-000 EBRO SEE ALSO MASSEY-FERGUSON

	CODE	MODEL	SCALE	MATER.	MANU.	STOCK #	CTRY.YEAR	REMARKS
☐	050118-001	6100	1/43	D	JOAL	250	S 82	WFE; Cab; Blue and gray or red and gray.
☐	050118-002	470	1/43	D	JOAL	103	S 81	WFE; Available with srpeader (#4); Yellow.

050120-000 EATON YALE SEE ALSO YALE-EATON

	CODE	MODEL	SCALE	MATER.	MANU.	STOCK #	CTRY.YEAR	REMARKS
☐	050120-001	TRACTOR SHOVEL	1/50	D	DINKY	973	GB 73	Industrial shovel; Articulated.

050315-000 ECONOMY

	CODE	MODEL	SCALE	MATER.	MANU.	STOCK #	CTRY.YEAR	REMARKS
☐	050315-001	15 HP	1/10	WOOD	KRUSE	NONE	USA 77	Gasoline engine.

050903-000 EICHER

	CODE	MODEL	SCALE	MATER.	MANU.	STOCK #	CTRY.YEAR	REMARKS
☐	050903-001A	KONIGSTIGER	1/18	D & P	MS TOY	1775	D ??	WFE; Gray; Promo.
☐	050903-001B	KONIGSTIGER	1/18	D & P	MS TOY	1775	D ??	WFE; Front-wheel-assist; Green and red; Hood raises.
☐	050903-001C	KONIGSTIGER	1/18	D & P	MS TOY	1875	D ??	WFE; Same except with front loader.
☐	050903-002A	GOODEARTH	1/20	D	MAXWELL	599	IN 80	WFE; Red; Represents a 1950's tractor
☐	050903-002B	GOODEARTH	1/20	D	MAXWELL	599	IN 83	WFE; Same except blue.
☐	050903-003A	3105A	1/35	D	CONRAD	01 3046	D ??	WFE.
☐	050903-003B	3105A	1/35	P	CONRAD	01 3046	D 80	WFE; Same except plastic.
☐	050903-004	KONIGSTIGER	1/15	P	MICHAEL SEIDEL		D 66	WFE; Gray.
☐	050903-005	1953	1/30	P	??		??	WFE; Promo.
☐	050903-006	1955	1/20	P	PENDING		??	WFE; Promo.
☐	050903-007	1960	1/30	P	PENDING		??	WFE; Promo.
☐	050903-008	??	20 cm	D	CURSOR		D ??	WFE.

051819-000 ERTL

	CODE	MODEL	SCALE	MATER.	MANU.	STOCK #	CTRY.YEAR	REMARKS
☐	051819-001	CLOD HOPPER	5"	D & P	ERTL	2501	USA 75	WFE; "Mod" type tractor.
☐	051819-002	GULLY WHUMPER	5"	D & P	ERTL	2502	USA 75	WFE; "Mod" type tractor.
☐	051819-003	OL McSMURF	2-1/2"	D & P	ERTL	1466	USA 83	WFE; Cartoon character figure on seat of tractor.

051903-000 ESCORT INDIAN VERSION OF FORD

	CODE	MODEL	SCALE	MATER.	MANU.	STOCK #	CTRY.YEAR	REMARKS
☐	051903-001	335	1/25	D	MORGAN MILTON LYD.		IN 77	WFE; Orange and white.
☐	051903-002	335	1/20	D	MAXWELL		IN 80	WFE; Orange and white; Headlights on sides of hood.

CODE	MODEL	SCALE	MATER.	MANU.	STOCK #	CTRY.	YEAR	REMARKS
☐ 051903-003	335	1/20	D	MAXWELL		IN	81	WFE; Orange and white; Headlights in the grill; Different seat and other detail.
☐ 051903-004	335	1/43	D	MATTELL-MINI		IN	84	WFE; Orange and white.

060108-000 FAHR

CODE	MODEL	SCALE	MATER.	MANU.	STOCK #	CTRY.	YEAR	REMARKS
☐ 060108-001	??	1/25	P	CURSOR		D	??	WFE.
☐ 060108-002	??	1/90	P	WIKING	38	D	56	WFE; Color variations; With vertical or horizontal exhaust stack.
☐ 060108-003	??	1/60	D	SIKU	V-48	D	??	WFE.
☐ 060108-004	??	1/50	P	CURSOR		D	55	WFE.
☐ 060108-005	??	1/32	P KIT	SIKU	293	D	??	WFE; Yellow; Represents a 1955 model.

060109-000 FAIRBANKS-MOORSE

CODE	MODEL	SCALE	MATER.	MANU.	STOCK #	CTRY.	YEAR	REMARKS
☐ 060109-001	Z-ENGINE	1/16	CI	ARCADE		USA	30	Model of small portable gasoline engine.
☐ 060109-002	Z-ENGINE	1/16	CA	OLD TIME TOYS: PIONEER TRACTOR WORKS	NONE	USA	68	Reproduction of Arcade engine.
☐ 060109-003	Z-ENGINE	1/10	CA	IRVIN	NONE	USA	??	Reproduction of 061315-001 by Arcade except not having the splash guard for the crank; Irvin name cast in.

060113-000 FAMULUS

CODE	MODEL	SCALE	MATER.	MANU.	STOCK #	CTRY.	YEAR	REMARKS
☐ 060113-001	??	1/87	P	ESPEWE		DDR	58	WFE; Blue or green; Rare.
☐ 060113-002	?/	1/16	P	PIKO		DDN	70	WFE; Battery electric motor; Red or blue; With a tipping trailer.
☐ 060113-003	??	1/20	P	G.B.Z.			??	WFE; Color variations.

060118-000 FARMTOY

CODE	MODEL	SCALE	MATER.	MANU.	STOCK #	CTRY.	YEAR	REMARKS
					INTERNATIONAL			
☐ 060118-001	FARMTOY 1206	1/16	D	SCALE MODELS		USA	81	Row crop; MODELS OF THE '60'S Series; Resembles an International.
☐ 060118-002	FARMTOY 806	1/16	D	SCALE MODELS		USA	81	Row crop; MODELS OF THE '60'S Series; Resembles an International Farmall.

060514-000 FENDT

CODE	MODEL	SCALE	MATER.	MANU.	STOCK #	CTRY.	YEAR	REMARKS
☐ 060514-001	?	1/25	P	CURSOR		D	??	WFE.
☐ 060514-002	F 250 GT	1/25	P	CURSOR		D	63	WFE; "Geretrager" with a bed on the front.
☐ 060514-003	F 250 GT	1/43	D	CURSOR	570	D	??	WFE; "Geretrager" with a bed on the front.
☐ 060514-004	??	1/43	D	CURSOR	067	D	??	WFE; Front-wheel-assist; Name "Veith" on tires.
☐ 060514-005	FAVORIT 4S TURBOMATIK	1/43	D	CURSOR	967	D	70	WFE; Front-wheel-assist; Cab.
☐ 060514-006	FAVORIT	1/43	D	CURSOR		D		WFE; Same except decals.
☐ 060514-007	FARMER 2	1/20	P	CURSOR		D	58	WFE; Green.
☐ 060514-008	FARMER 2	1/25	P	CURSOR		D	63	WFE; Green and gray.
☐ 060514-009A	TURBO. FARMER 308LS	1/16	P	BRUDER	8811	D	84	WFE; Front-wheel-assist; Bright green; Either 308LS or 309LS decals.
☐ 060514-009B	TURBO. FARMER 308LS	1/16	P	BRUDER	8812	D	84	WFE; Front-wheel-assist; Cab; With loader; Dull green.
☐ 060514-009C	TURBO. FARMER 308LS	1/16	P	BRUDER	8813	D	84	WFE; Front-wheel-assist; Cab; With side mounted mower.
☐ 060514-010	FAVORIT 620LS TURBO.	1/32	D	GAMA	2307	D	81	WFE.
☐ 060514-011A	DIESELROSS 6 PS	1/35	D	CURSOR	780	D	80	WFE; Represents a 1930 model.
☐ 060514-011B	DIESELROSS 6 PS	1/32	D	CURSOR		D	81	WFE; Same except with mounted plow in lucite display case; Only 500 made.
☐ 060514-012	TURBOMATIC FAVORIT	4"	D	CURSOR	677	D	78	WFE; Front-wheel-assist; Cab; All green.
☐ 060514-013	FAVORIT 612SL	1/41	D & P	POLISTIL	CE-114	I	80	WFE; Front-wheel-assist; Cab.
☐ 060514-014	F255GT	9"	P	BRUDER		D	81	WFE; "Geretrager" with front mounted bed; Some variations also had a side mounted mower.
☐ 060514-015	FAVORIT TURBOMATIK		C KIT	BIJAGE VAN HET		NL	81	WFE.
☐ 060514-016	TURBO. FARMER 308LS	1/43	D	YAXON	040	I	82	WFE; Front-wheel-assist; Half cab.
☐ 060514-017	FAVORIT 44	1/25	D	CURSOR		D	70	WFE; No cab.
☐ 060514-018	FAVORIT FW-140	1/25	P	STRENCO	802	D	60	WFE.
☐ 060514-019	FARMER 309LS TURBO	1/32	D	SIKU		D	84	WFE; Front-wheel-assist; Cab.
☐ 060514-020	FARMER	10"	P	DBGM		D	84	WFE; Green, red and white; With a wagon.
☐ 060514-021A	FARMER	1/12	P	KARKURO		D	??	WFE.
☐ 060514-021B	FARMER	1/12	P	KARKURO		D	??	WFE; Same with front loader.
☐ 060514-022A	F 275 GT	1/43	D	CURSOR	478	D	??	WFE; Tool carrier.
☐ 060514-022B	F 250 GT	1/43	D	CURSOR	478	D	??	WFE; Tool carrier; Industrial orange.

060518-000 FERGUSON

CODE	MODEL	SCALE	MATER.	MANU.	STOCK #	CTRY.	YEAR	REMARKS
☐ 060518-001A	(30)	1/43	D	BENBROS		GB	??	WFE; Driver; Available with log trailer.
☐ 060518-001B	(30)	1/43	D	BENBROS		GB	??	WFE; Industrial model with front loader; Yellow.

CODE	MODEL	SCALE	MATER.	MANU.	STOCK #	CTRY.YEAR	REMARKS
☐ 060518-002	(30)	3-1/2"	D	MOKO-LESNEY		GB 51	WFE.
☐ 060518-003A	(30)	1/16	D	CHAD VALLEY		GB 55	WFE; Steerable; Opening hood; Excellent detail; Green.
☐ 060518-003B	(30)	1/16	D	CHAD VALLEY		GB 55	WFE; Same except gray and has a three-point-hitch; Promo.
☐ 060518-004	(30)	1/13	P	TOPPING MODELS		USA 54	WFE; Three-point-hitch; Gray; Available with a three-point-hitch disk plow.
☐ 060518-005	(20)	1/12	POT	ADVANCED PRODUCTS		USA 48	WFE; Gray; Three-point-hitch; Some had bodies made of a non-metallic material; Wheel variations include a steel rim with a rubber tire or a hard rubber wheel/tire.
☐ 060518-006A	(30)	1/43	D	TEKNO	460	DK 54	WFE; Red; Hood opens; Available with implements; Decal diagonally across top of hood.
☐ 060518-006B	(30)	1/43	D	TEKNO	460	DK 54	WFE; Same except gray; Rare.
☐ 060518-007	(TE-20)	5-1/2"	D	METTOY-CASTOYS		GB ??	WFE; Steerable; Hood opens; With or without clockwork; Color variations; Available with a very well detailed plow.
☐ 060518-008	(TE-30)	1/20	PK	AIRFIX		GB 50	WFE; Color variations.
☐ 060518-009	30	1/25	P	ALLEMAGNE-PLASTY	8	D 55	WFE.
☐ 060518-010	(30)	4-1/2"	D	FUN-HO	402	NZ ??	WFE; Color variations.
☐ 060518-011	(35)	6-1/2"	D	FUN-HO	520	NZ ??	WFE; Color variations.
☐ 060518-012	(35)	1/32	D	MICRO-MODELS	4337	NZ 55	WFE.
☐ 060518-013	30	1/16	P	NOVA		F 51	WFE; Promotional model.
☐ 060518-014	20 DEMONSTRA-TOR	1/10	D	MILLS BROS.		GB 50	WFE; This tractor demonstrator came in a special container which could be converted into a "track" to demonstrate the difference in pulling a trailer type plow and a three-point-hitch plow. Ferguson salesmen used the kit to sell the virtues of the "Ferguson System".
☐ 060518-015	TE-20	1/35	D	CONRAD	3405	D 81	WFE; Represents a 1950 model; Available with a three-point-hitch mounted plow.
☐ 060518-016A	FERGUSON-BLACK	1/32	WM KIT	BROWN'S MODELS		GB 82	WFE; Represented the first Ferguson prototype.
☐ 060518-016B	FERGUSON-BLACK	1/32	WM KIT	BROWN'S MODELS		GB 82	WFE; "Steel" wheels; Gray; This kit was available with rubber tires although the real Ferguson-Black never had rubber tires.
☐ 060518-017	(TO-30)	1/20	PK	THOMAS TOYS		USA ??	WFE.
☐ 060518-018	TE-20	1/20	P	TRITON (LEGO)		DK 66	WFE; Available with a variety of implements.
☐ 060518-019	TE-20	1/43	D	REPLICA DIE CAST		AUS 84	WFE; Copy of 130106-002A Massey-Ferguson.
☐ 060518-020	(30)		D	CRESCENT		GB ??	WFE; See also MF 13016-005.

060901-000   FIAT

CODE	MODEL	SCALE	MATER.	MANU.	STOCK #	CTRY.YEAR	REMARKS
☐ 060901-001	FIAT CONCORD 700-S	1/43	D	BUBY	1038	RA 71	WFE; Model of Fiat made in Argentina.
☐ 060901-002A	550	1/35	D	DUGU	2	I 66	WFE.
☐ 060901-002B	600	1/35	D	DUGU	2	I ??	WFE.
☐ 060901-002C	640	1/35	D	DUGU	2	I ??	WFE.
☐ 060901-003A	25R	1/43	D	MINI-GAMA	914	D 61	WFE; Orange with white wheels.
☐ 060901-003B	25R	1/43	D	MINI-GAMA	930	D 61	WFE; With blade.
☐ 060901-003C	25R	1/43	D	MINI-GAMA	914	D 61	WFE; With trailer.
☐ 060901-003D	25R	1/43	D	MINI-GAMA	940	D 61	WFE; Same with loader.
☐ 060901-004A	550	1/36	D	ZISS R W MODELLE		D 73	WFE.
☐ 060901-004B	600	1/36	D	ZISS R W MODELLE		D 73	WFE.
☐ 060901-005A	780	1/43	D	FORMA-PLAST	055	I 76	WFE; Poor quality wheels and tires which "melt" together.
☐ 060901-005B	8BODT	1/43	D	FORMA-PLAST	056	I 76	WFE; Front-wheel-assist; Poor quality wheels and tires which "melt" together.
☐ 060901-006A	40 CA	1/60	D	SIKU	V-238	D 64	Crawler with loader.
☐ 060901-006B	40 CA	1/60	D	SIKU	V-239	D 64	Crawler with blade.
☐ 060901-007A	FL-10	1/24	D	MINI-AUTO	6032	D 73	Crawler with loader; Excellent detail.
☐ 060901-007B	FIAT-ALLIS FL-10	1/24	D	MINI-AUTO		D ??	Crawler with loader; Excellent detail; See also Allis-Chalmers.
☐ 060901-008	640	1/36	D & P	OLD CARS	51	I 78	WFE; Replaces Dugu models; With or without 640 decals.
☐ 060901-009	FIAT-ALLIS 41-B	1/50	D	GESCHA		D 79	Crawler with blade and ripper; Cab.
☐ 060901-010A	8BODT	1/32	D & P	BRITAINS LTD.	9528	GB 79	WFE; Front-wheel-assist; Rear lift hitch; Orange.
☐ 060901-010B	HESSTON 8BODT	1/32	D & P	BRITAINS LTD.	9528	GB 81	WFE; Same except environmental red; 2000 made. See also Hesston.

CODE	MODEL	SCALE	MATER.	MANU.	STOCK #	CTRY.	YEAR	REMARKS
060901-010C	8BODT	1/32	D & P	BRITAINS LTD	5309	GB	83	WFE; Same except with snow plow; Sold only in Italy in this combination.
060901-010D	880/HALF TRACKS	1/32	D & P	BRITAINS LTD	9527	GB	82	WFE; Front-wheel-assist; Environmental red; With halftracks.
060901-010E	880 DT	1/32	D & P	BRITAINS LTD	9527	GB	82	WFE; Same except without halftracks.
060901-011A	780	1/13	P	FORMA-PLAST		I	78	WFE; Excellent detail.
060901-011B	880 DT	1/13	P	FORMA-PLAST		I	78	WFE; Front-wheel-assist; Excellent detail.
060901-012A	(780)	1/87	P	MERCURY	810	I	79	WFE; White cab.
060901-012B	(780)	1/87	P	MERCURY	810	I	??	WFE; Orange; Cab.
060901-013	DIM 30		D	OLD CARS	52	I	80	Industrial forklift.
060901-014A	(25R)	1/42	D	BUDGIE	306	GB	58	WFE; Cab; Loader; Represents a 1951 model.
060901-014B	(25R)	1/42	D	BUDGIE	314	GB	??	WFE; Same except with blade.
060901-015	880	1/19	P	G.P.	102		??	WFE; Yellow and black or red and orange.
060901-016	880 DT	1/43	D	YAXON		I	80	WFE; Front-wheel-assist; Identical to Forma-Plast but marked Yaxon.
060901-017	880 DT	1/43	D	YAXON	2501	I	81	WFE; Front-wheel-assist; Same except sold by Mattell.
060901-018	880 DT	1/43	D	YAXON S.A.		S.A.	80	WFE; Front-wheel-assist; Orange wheels; S.A. on box only.
060901-019	FIAT-ALLIS FL-20	1/50	D	OLD CARS		I	??	Industrial crawler.
060901-020	FIAT-ALLIS 41-B	1/50	D	CONRAD		D	??	Industrial crawler.
060901-021	702 50TH ANN.	1/12	P	MOPLAS		I	69	WFE; "Steel" wheels; Represents a 1919 model; Commomerative 50th Anniversary model.
060901-022	44-23	1/16	SC	SCALE MODELS		USA	82	WFE; Front-wheel-drive; Articulated; Cab; Italian version of the Versatile 850.
060901-023	FIAT-ALLIS FR20	1/50	D	OLD CARS		I	82	Industrial four-wheel-loader.
060901-024	880 DT-5	1/32	D & P	YAXON		I	83	WFE; Front-wheel-assist; Cab; Also sold in a promo box.
060901-025	FORKLIFT	1/50	D	OLD CARS		I	83	Industrial forklift.
060901-026	FIAT-ALLIS FE—40	1/50	D	OLD CARS		I	83	Industrial excavator.
060901-027	(550)	1/50	P	GRISONI GIOCATTOLI		I	78	WFE; Red; Also available with blade or front loader.
060901-028	BD 20	1/20	P	ITES		CS	??	Crawler; Either with loader or blade; Red or yellow.
060901-029	(550DT)	1/10	P	COMA	5156	I	??	WFE; Front-wheel-assist; Crude plastic model.
060901-030	(550)	1/10	P	COMA-INTER-TOYS	5156	I	??	WFE; Crude plastic model.
060901-031	(880)	1/10	P	RENAM		I	??	WFE; Crude plastic model; Yellow.
060901-032	(880)	1/16	P	DICKIE (SPIEIZEUG)	2930	D	82	WFE; Remote control; With trailer.
060901-033	640	1/25	D	TTF		T	??	WFE; Reported to have been made in the Turkish Fiat factory; Crude.
060901-034	180-90 TURBO	1/28	P	GIODI		I	85	WFE; Cab.

061301-000 FIELD MARSHALL

CODE	MODEL	SCALE	MATER.	MANU.	STOCK #	CTRY.	YEAR	REMARKS
061301-001A	SERIES I	1/43	D	DINKY	27N	GB	52	WFE; Metal tires/wheels; Tan driver; Color variations.
061301-001B	SERIES I	1/43	D	DINKY	301	GB	54	WFE; Rubber tires; Painted driver; One cylinder engine.
061301-002	SERIES II	1/32	WM KIT	BROWN'S MODELS		GB	79	WFE; Represents a 1948 model.
061301-003	SERIES III	1/12	CUS.	KARSLAKE	NONE	GB	82	WFE; Also a Series IIIA made; Limited production.
061301-004	SERIES II	1/32	WM KIT	BRIAN PARKS	T14	GB	84	WFE; Green.

061504-000 FODEN

CODE	MODEL	SCALE	MATER.	MANU.	STOCK #	CTRY.	YEAR	REMARKS
061504-001	??	1/43	CUS.	PHANTOM MODELS	A01	F	??	WFE; Model of 1896 steam traction engine.

061718-000 FORD

CODE	MODEL	SCALE	MATER.	MANU.	STOCK #	CTRY.	YEAR	REMARKS
061718-001	(9N)	6-1/2"	CI	ARCADE		USA	40	WFE; Cast in driver; Some models have an attached three-point-hitch plow.
061718-002	(9N)	6-1/2"	CI	ARCADE		USA	41	WFE; Cast in driver; With mounted rear dump earth mover.
061718-003	(9N ROW CROP)	3-1/4"	CI	ARCADE		USA	40	Row crop; Cast in driver; Ford never made a 9N with the row crop style front end.
061718-004	8N	1/12	P	PRODUCT MINIATURE		USA	52	WFE; Three-point-hitch; The real Ford 8N was introduced in 1947.
061718-005A	NAA JUBILEE	1/12	P	PRODUCT MINIATURE		USA	53	WFE; Three-point-hitch; No headlights; "GOLDEN JUBILEE" model.
061718-005B	600	1/12	P	PRODUCT MINIATURE		USA	55	WFE; Three-point-hitch; Headlights.
061718-006	900	1/12	P	PRODUCT MINIATURE		USA	55	Row crop; Three-point-hitch.
061718-007	(8N)	1/12	P	ALUM. METAL PROD. (AMT)		USA	50	WFE; Clockwork.
061718-008	(8N)	1/12	P	M.P.C.		USA	52	WFE; Clockwork; Long body.
061718-009A	(8N)	5"	CA	QUIRALU		F	47	WFE; Separately cast driver; Clockwork; Color variations.

	CODE	MODEL	SCALE	MATER.	MANU.	STOCK #	CTRY.	YEAR	REMARKS
☐	061718-009B	(8N)	5"	CA	QUIRALU		F	50	WFE; Same except with clockwork.
☐	061718-010A	(8N)	1/32	D	TOOTSIETOY	290	USA	??	WFE; Red; With loader.
☐	061718-010B	(8N)	1/32	D	TOOTSIETOY		USA	??	WFE; Red and silver.
☐	061718-011	(NAA JUBILEE)	1/32	D	TOOTSIETOY		USA	??	WFE; Red.
☐	061718-012A	(NAA ROAD—MASTER)	1/32	D	DCMT LONE-STAR	1258	GB	??	WFE; "FARM KING" decals; Red, yellow or blue; Similar to Tootsietoy Ford Jubilee.
☐	061718-012B	(NAA ROAD-MASTER)	1/32	D	DCMT LONE-STAR		GB	??	WFE; Same except with cab.
☐	061718-013	(8N)	1/24	P	HUBLEY	309	USA	??	WFE; Gray.
☐	061718-014	(960) H	1/10	D	HUBLEY	90	USA	??	Row crop; Red and gray; Available with implements.
☐	061718-015A	961 POWERMASTER	1/12	D	HUBLEY	507	USA	61	WFE; Three-point-hitch; Red and gray.
☐	061718-015B	961 POWERMASTER	1/12	D	HUBLEY	507	USA	61	Row crop; Three-point-hitch; Red and gray.
☐	061718-015C	SELECT-O-SPEED	1/12	D	HUBLEY	508	USA	62	Row crop; Three-point-hitch; Red and gray; No decals.
☐	061718-016A	(4000)	1/12	D	HUBLEY	1508	USA	65	Row crop; Three-point-hitch; Blue and gray; Name cast in grill.
☐	061718-016B	4000	1/12	D	HUBLEY	1508	USA	65	WFE; Three-point-hitch; Blue and gray; Name cast in grill.
☐	061718-017A	6000	1/12	D	HUBLEY	509	USA	63	Row crop; Chromed vertical exhaust; Red and gray.
☐	061718-017B	6000	1/12	D	HUBLEY	509	USA	64	Row crop; Same except blue and gray.
☐	061718-018A	COMMANDER 6000	1/12	D	HUBLEY	1509	USA	63	Row crop; Plastic exhaust and air cleaner; Three-point-hitch; Blue and gray.
☐	061718-018B	COMMANDER 6000	1/12	D	HUBLEY-GABRIEL	26157	USA	79	Row crop; Same except no exhaust, air cleaner or three-point-hitch.
☐	061718-019	1841 INDUSTRIAL	1/12	T	CRAGSTAN	90010	J	??	WFE; Loader-backhoe; Remote control.
☐	061718-020	4040	1/12	T	CRAGSTAN		J	??	WFE; No. 21 forklift on rear; Remote control.
☐	061718-021	4040	1/12	T	CRAGSTAN		J	??	WFE; Red; Backhoe/loader; Remote control.
☐	061718-022	4040 HD	1/12	T	CRAGSTAN		J	??	WFE; Backhoe/loader; Remote control; Yellow.
☐	061718-023A	(4000)	1/56	D	LESNEY	39	GB	67	WFE; blue.
☐	061718-023B	(4000)	1/56	D	LESNEY	39	GB	67	WFE; Yellow over blue.
☐	061718-023C	(4000)	1/56	D	LESNEY	39	GB	69	WFE; Orange; Available as a set consisting of a truck tractor with flat bed trailer and three tractors; Rare.
☐	061718-024	(4000)	1/56	D	GUISVAL	250	E	??	WFE; Separate driver; Green and yellow; Similar to Lesney Ford 061718-023.
☐	061718-025	4000	1/12	D	ERTL	805	USA	65	WFE; Three-point-hitch; Two section grill.
☐	061718-026A	4000	1/12	D	ERTL	805	USA	68	WFE; Three-point-hitch; One section grill.
☐	061718-026B	4400 INDUSTRIAL	1/12	D	ERTL	805	USA	72	WFE; Three-point-hitch; One section grill; Yellow.
☐	061718-027A	4600	1/12	D	ERTL	A-805	USA	77	WFE; Same except no three-point-hitch.
☐	061718-027B	4600	1/12	D	ERTL	A-805	USA	76	WFE; Three-point-hitch; Flat fenders.
☐	061718-028A	5550 IND.	1/12	D	ERTL	820	USA	72	WFE; Industrial backhoe-loader; Yellow.
☐	061718-028B	7500 IND.	1/12	D	ERTL	820	USA	75	WFE; Same except decals.
☐	061718-028C	750 IND.	1/12	D	ERTL	820	USA	80	WFE; Same except decals.
☐	061718-029A	8000	1/12	D	ERTL	800	USA	68	WFE; Three-point-hitch; Small "8000" decals.
☐	061718-029B	8000	1/12	D	ERTL	800	USA	70	WFE; Large "8000" decals; With or without three-point-hitch.
☐	061718-029C	8600	1/12	D	ERTL	A-800	USA	73	WFE; With or without three-point-hitch.
☐	061718-029D	9600	1/12	D	ERTL	A-821	USA	74	WFE; Dual wheels; With or without three-point-hitch.
☐	061718-030	145 HYDRO LGT	1/12	D	ERTL	808	USA	72	Lawn and garden tractor with two-wheel trailer.
☐	061718-031	(4000)	7"	D & P	GAMA	1856	D	??	WFE; Remote control; No. 1866 has four-wheel wagon.
☐	061718-032	(4000)	4-1/2"	D	GAMA		D	??	WFE; Clockwork; Red, yellow and green.
☐	061718-033	(4000)	4-1/4"	D & T	GAMA		D	??	WFE; Clockwork.
☐	061718-034A	(4000)	1/43	D	GAMA	9141	D	??	WFE; Turf type tires; Red, green and yellow.
☐	061718-034B	(4000)	1/43	D	GAMA	9341	D	??	WFE; Same except with blade.
☐	061718-034C	(4000)	1/43	D	GAMA	9401	D	??	WFE; Same except with front loader.
☐	061718-035	(4000)	2-1/4"	D	FUN-HO		NZ	??	WFE; Cast in driver.
☐	061718-036	(4000)	3-3/4"	CA	FUN-HO	103	NZ	??	WFE; Cast in driver.
☐	061718-037	5000 SUPER MAJOR	1/32	D	BRITAINS LTD.	9630	GB	??	WFE; Cab; Industrial yellow; With Shawnee Pool gooseneck dump trailer.
☐	061718-038A	5000	1/32	D	BRITAINS LTD	9527	GB	69	WFE; Gary; Rounded cab.

CODE	MODEL	SCALE	MATER.	MANU.	STOCK #	CTRY.	YEAR	REMARKS
061718-038B	5000	1/32	D	BRITAINS LTD.	9527	GB	71	WFE; Blue; Square cab.
061718-038C	5000	1/32	D	BRITAINS LTD.	9527	GB	71	WFE; No cab.
061718-039	5000	1/43	PK	BRITAINS LTD.	1101	GB	??	WFE; Snap together plastic kit; "Mini Series".
061718-040A	6600	1/32	D	BRITAINS LTD.	9524	GB	76	WFE; Square cab; Larger, oval muffler; Front wheel variations.
061718-040B	6600/ROPS & SCRAPER	1/32	D	BRITAINS LTD.	9559	GB	85	WFE; With ROPS and rear mounted yard scraper.
061718-041A	5000 SUPER MAJOR	1/43	D	CORGI	67	GB	66	WFE.
061718-041B	5000 SUPER MAJOR	1/43	D	CORGI	72	GB	70	WFE; With rear trencher.
061718-041C	5000 SUPER MAJOR	1/43	D	CORGI	74	GB	70	WFE; With side trencher.
061718-041D	5000 SUPER MAJOR	1/43	D	CORGI	67	GB	70	WFE; Same except with simplified lift.
061718-042A	5000	1/55	D	MAJORETTE	255	F	72	WFE; Red or metallic blue; Available with a variety of implements.
061718-042B	5000	1/55	D	MAJORETTE DE BRASIL	255	BR	??	WFE; Same except made in Brazil.
061718-043	(4000)	1/61	D	TINY CAR		BR	68	WFE.
061718-044	(4000)	10"	P	GAY TOYS		USA	75	WFE; Color variations; Also available with front loader.
061718-045	(145) LGT		P	GAY TOYS	615	USA	75	Lawn and garden tractor; Color variations.
061718-046	8600	8-1/2"	P	PROC. PLASTIC PROD.		USA	76	WFE; Cab; Chromed engine; Color variations; Decal variations.
061718-047	4550 IND.	1/35	D	NZG	130	D	72	WFE; Industrial backhoe-loader.
061718-048	(9N)	1/12	CA	TOY & TACKLE		USA	??	WFE; With cast in driver; Three-point-hitch; Replica of Arcade (9N) 061718-001.
061718-049	(4000)	1/43	P	GALANITE		S	76	WFE; Molded in driver.
061718-050	7700	1/12	D	ERTL	819	USA	77	WFE; Cab.
061718-051	9700	1/12	D	ERTL	817	USA	77	WFE; Cab; Duals.
061718-052A	9700	1/64	D	ERTL	1621	USA	78	WFE; Cab; Wheel variations.
061718-052B	TW-20	1/64	D	ERTL	1621	USA	??	WFE; Cab; Wheel and grill decal variations.
061718-053	3600	1/20	D	MORGAN MILTON LTD.		IN	67	WFE; Air cleaner stack.
061718-054	2N	1/12	KOR	PIONEERS OF POWER		USA	78	WFE; "Steel" wheels; Replica of Arcade Ford 061618-001; Represents a 1939 model.
061718-055	(NAA JUBILEE)	5-1/2"	P	WYANDOTTE		USA	??	WFE.
061718-056	(4000)	7"	D	?			??	WFE; Cast in two sections.
061718-057	5000 SUPER MAJOR	1/43	D	MAXWELL		IN	80	WFE; Copy of Corgi 5000; Variety of colors.
061718-058	5000 SUPER MAJOR		P	HOVER	416A	HK	??	WFE; Battery operated; Orange and blue.
061718-059	5000 SUPER MAJOR		P	HOVER	853	HK	??	WFE; Red and white.
061718-060	5000		P	QUELLE INTER.	318-8430		??	WFE; With disk (153) or cattle trailer (271).
061718-061	5000 SUPER MAJOR	3"	P	TIMPO		GB	??	WFE; Red and blue with black grill.
061718-062	6600	4-1/4"		IKE TOYS		DK	??	WFE; With two-wheel trailer.
061718-063	(6600)	1/43	D	YAXON	069	I	??	WFE; Front-wheel-assist; Cab; With plow 0146.
061718-064	8700	1/41	P	POLISTIL	CE115	I	80	WFE; Front-wheel-assist; Half cab.
061718-065	9700 PULLER	1/32	PK	IMAI	B-972	J	81	WFE; Battery electric motor.
061718-066A	TW-10	1/12	D	ERTL	819	USA	81	WFE; Cab.
061718-066B	TW-20	1/12	D	ERTL	819	USA	82	WFE; Cab.
061718-066C	TW-25	1/12	D	ERTL	819	USA	83	WFE; Cab.
061718-066D	TW-35	1/12	D	ERTL	819	USA	83	WFE; Cab; Duals.
061718-067A	TW-20	1/32	D	BRITAINS LTD.	9523	GB	81	WFE; Front-wheel-assist; Cab; Front weights.
061718-067B	7710	1/32	D	BRITAINS LTD.	9523	GB	83	WFE; Same except decals.
061718-067C	7710 IND.	1/32	D	BRITAINS LTD.	9812	GB	84	WFE; Same except industrial yellow; "Autoway Series".
061718-068	(6600)	1/32	D	BRITAINS LTD.	9420	GB	82	WFE; Wrong colors—green or gray; No decals.
061718-069	TW-20	1/32	D	ERTL	1643	USA	81	WFE; Cab.
061718-070A	FW-60	1/32	D	ERTL	1926	USA	80	WFE; Four-wheel-drive; "TOY FARMER CLASSIC"; 1000 made.
061718-070B	FW-60	1/32	D	ERTL	1926	USA	81	WFE; Four-wheel-drive; Cab; Articulated.
061718-071	FW-60	1/32	P	ERTL		USA	84	WFE; Four-wheel-drive; Articulated; Cab; "TOY FARMER" special edition issue; "Classic Collectors Series" molded in.
061718-072A	FW-60	1/64	D	ERTL	1528	USA	82	WFE; Four-wheel-drive; Articulated; Gray wheels; Cab.
061718-072B	FW-60	1/64	D	ERTL	1528	USA	83	WFE; Four-wheel-drive; Articulated; White wheels; Cab.
061718-073	4000	1/12	D	SCALE MODELS		USA	82	Row crop; Reissue of Hubley Ford 4000.
061718-074	555 IND.	1/32	D	ERTL	1456	USA	82	WFE; Industrial backhoe-loader.
061718-075A	7710	1/16	D	ERTL	836	USA	82	WFE; Front-wheel-assist; ROPS.
061718-075B	7710 COLL. CLASSIC	1/16	D	ERTL		USA	83	WFE; Front-wheel-assist; ROPS; Collector Classic.

	CODE	MODEL	SCALE	MATER.	MANU.	STOCK #	CTRY.	YEAR	REMARKS
☐	061718-075C	7710 TOY FARMER	1/16	D	ERTL		USA	83	WFE; Front-wheel-assist; ROPS; Show tractor.
☐	061718-076	CL-25 LD. HAND.	1/16	D	ERTL	840	USA	83	Skid steer loader.
☐	061718-077	TW-15	1/32	P	ERTL	819	USA	83	WFE; Cab; Radio controlled.
☐	061718-078	5000	8"	P	HOYUK		TU	83	WFE; Molded in driver; With rake.
☐	061718-079	BIG BLUE	750 ML	PORC.	PACESETTER	NO. 3	USA	83	WFE; Porcelain and plastic decanter.
☐	061718-080	BIG BLUE	50 ML	PORC.	PACESETTER	NO. 3	USA	83	WFE; Porcelain and plastic decanter.
☐	061718-081	(4000) PULLER	1/12	P	GAY TOYS	740	USA	83	WFE; "Mod" puller tractor.
☐	061718-082	3600	1/43	D	MATTELL MINI	1001	IN	84	WFE.
☐	061718-083	3600	1/20	D	MAXWELL		IN	80	WFE; Air cleaner stack.
☐	061718-084	8210	1/43	D	YAXON	045	I	84	WFE; Front-wheel-assist; Cab.
☐	061718-085	FW-60	1/16	CUS.	DAVID SHARP	NONE	USA	79	WFE; Four-wheel-drive; Articulated; Cab.
☐	061718-086A	LGT 12-LIM. ED.	1/12	D	ERTL	808TA	USA	84	Lawn and garden tractor; Limited Edition "SPECIAL EDITION MARCH 84": molded into operator's platform.
☐	061718-086B	LGT 12	1/12	D	ERTL	808	USA	84	Lawn and garden tractor.
☐	061718-087	(5000)	1/48	PK	ESCI	4025	I	78	WFE; Military airport tractor; Part of a set of aircraft support group.
☐	061718-088	(8N)	1/10	D	MILLS BROS. ?		??		WFE; Similar to the Ferguson 30 in the Mills Brothers demonstration set 060518-014.
☐	061718-089	(8000)	1/64	PEWTER	HALLMARK	C Series 2	USA	80	WFE; With wagon.
☐	061718-090	(8N)	1/24	P	SCALE MODELS		USA	84	WFE; Re-issue of Hubley (061718-013).
☐	061718-091	7610	1/32	D	LONE STAR		GB	84	WFE; Cab.
☐	061718-092	(JUBILEE NAA)	1/32	D	NOSTALIG		USA	84	WFE; Represents a 1953 Golden Jubilee model.
☐	061718-093	BN/LOADER	1/43	D	ERTL	2512	USA	85	WFE; With loader; Represents a 1947 model.

061719-000 FORDSON

	CODE	MODEL	SCALE	MATER.	MANU.	STOCK #	CTRY.	YEAR	REMARKS
☐	061719-001	(F)	3-7/8"	CI	ARCADE	273X	USA	26	WFE; Cast in driver; "Steel" wheels or rubber tires; Color variations.
☐	061719-002	(F)	4-3/4"	CI	ARCADE	274X	USA	26	WFE; Cast in driver; "Steel" wheels or rubber tires; Color variations.
☐	061719-003	(F)	5-3/4"	CI	ARCADE	280X	USA	32	WFE; Cast in driver; "Steel" wheels or rubber tires; Color variations.
☐	061719-004	(F)	6"	CI	ARCADE	275X	USA	32	WFE; Separately cast, plated driver; Smooth or lugged "steel" wheels; Color variations.
☐	061719-005	(F)	6"	CI	ARCADE		USA	??	WFE; Separately cast, plated driver; "W & K" solid cast wheels with thin rubber industrial type tires; Color variations.
☐	061719-006	(F)	3-3/4"	CI	HUBLEY		USA	28	WFE; Cast in driver; Color variations.
☐	061719-007	(F)	3-1/2"	CI	HUBLEY	336	USA	38	WFE; Cast in driver; Rubber wheels/tires; Color variations.
☐	061719-008	(F)/LOADER	8-1/2"	CI	HUBLEY	727	USA	38	WFE; Separately cast; Plated driver; "Steel" wheels or rubber wheels/tires; Loader operated with side crank.
☐	061719-009	(F)	5-1/2"	CI	HUBLEY		USA	??	WFE; "Steel" wheels or rubber tires/wheels.
☐	061719-010	(F)	4"	CI	KILGORE		USA	??	WFE; "Steel" wheels or rubber tires/wheels; Cast in driver.
☐	061719-011	(F)	5-1/8"	CI	KILGORE		USA	??	WFE; "Steel" wheels; Cast in driver.
☐	061719-012	(F)	6-1/2"	CI	KENTON		USA	??	WFE; "Steel" wheels; Separately cast driver; Has "water washer" air cleaner cast in just behind the engine.
☐	061719-013	(F)	3-1/2"	CI	NORTH & JUDD		S	??	WFE; Separately cast, plated driver. "Steel" wheels.
☐	061719-014	(F)	5-3/4"	CI	DENT		USA	??	WFE; Separately cast, plated driver; "Steel" wheels; Fewer hood bands.
☐	061719-015A	(F)	1/43	D	DINKY	22E	GB	34	WFE; "Steel" wheels; Fenders; Marked "Hornby".
☐	061719-015B	(F)	1/43	D	DINKY	22E	GB	36	WFE; "Steel" wheels; Fenders; Marked "Dinky".
☐	061719-016	(F)	1/43	D CAT	VARNEY COPY	1	GB	76	WFE; Copy of Dinky (061719-015).
☐	061719-017	(F)	6"	CA	OTT, PTW		USA	69	WFE; Reproduction of Arcade Fordson, "Steel" wheels.
☐	061719-018	(F)	1/16	D	ERTL	850	USA	69	WFE; "Steel" wheels; Steerable; No driver.
☐	061719-019	(F)	5-1/2"	D	FUN-HO	104-B	NZ	??	WFE; Cast in driver; "Steel" wheels.
☐	061719-020	SUPER MAJOR	1/87	D	FUN-HO	16	NZ	67	WFE.
☐	061719-021	(F)	1/16	KOR	CHARLES SOUHRADA	NONE	USA	75	WFE; American or English color versions; "Steel" wheels; Driver; Fenders.
☐	061719-022A	(F) FARM	1/87	PK	JORDAN PRODUCTS	C-218	USA	75	WFE; "Steel" cleated wheels.
☐	061719-022B	(F) INDUSTRIAL	1/87	PK	JORDON PRODUCTS	C-219	USA	75	WFE; Smooth solid wheels.
☐	061719-023	(F)	9-1/4"	PORC	EZRA BROOKS		USA	72	WFE; Glass whiskey bottle.

	CODE	MODEL	SCALE	MATER.	MANU.	STOCK #	CTRY.	YEAR	REMARKS
☐	061719-024	(F)	3-3/4"	CI	A.C. WILLIAMS		USA	??	WFE; Cast in driver; Plated "Steel" wheels or rubber tires/wheels; Color variations.
☐	061719-025	(F)	4-3/4"	CI	A.C. WILLIAMS		USA	??	WFE; Cast in driver; Plated "Steel" wheels or rubber tires/wheels; Color variations.
☐	061719-026	(F)	5-3/4"	CI	A.C. WILLIAMS		USA	??	WFE; Cast in driver; Plated "Steel" wheels or rubber tires/wheels; Color variations.
☐	061719-027	(F)	4-1/2"	CI	A.C. WILLIAMS		USA	??	Roller front instead of wheels.
☐	061719-028	(E27N)	1/76	D	BRITAINS LTD.	LV-604	GB	53	WFE; With or without driver.
☐	061719-029A	(E27N)	1/32	D	BRITAINS LTD.	127F	GB	48	WFE; "Steel" wheels; Those made from '48 to '53 have a tapered steering column while those made between '53 and '57 have a straight steering column.
☐	061719-029B	(E27N)	1/32	D	BRITAINS LTD.	128F	GB	48	WFE; Rubber tires; Those made from '48 to '53 have a tapered steering column whle those made between '53 to '57 have a straight steering column.
☐	061719-030A	POWER MAJOR	1/32	D	BRITAINS LTD.	171F	GB	59	WFE; "Steel" wheels; Headlights on the sides of the hood.
☐	061719-030B	POWER MAJOR	1/32	D	BRITAINS LTD.	172F-9525	GB	59	WFE; Rubber tires; Headlights on the sides of the hood.
☐	061719-031A	SUPER MAJOR	1/32	D	BRITAINS LTD.	171F	GB	61	WFE; "Steel" wheels; Headlights in the grill; Small triangle cast on sides of hood.
☐	061719-031B	SUPER MAJOR	1/32	D	BRITAINS LTD.	172F-9525	GB	63	WFE; Rubber tires; Headlights in the grill; Fordson imperial blue; Either Ford or Fordson style rear wheels; The later version has a light cat into the rear fender.
☐	061719-031C	INDUSTRIAL MAJOR	1/32	D	BRITAINS LTD.	9630	GB	63	WFE; Rubber tires; Cab; Yellow; Light cast onto rear fenders.
☐	061719-032A	5000 SUPER MAJOR	1/32	D	BRITAINS LTD.	9527	GB	65	WFE.
☐	061719-032B	5000 SUPER MAJOR	1/32	D	BRITAINS LTD.	9527	GB	71	WFE; Early style rounded cab.
☐	061719-033	5000	1/32	D	BRITAINS LTD.	9527	GB	73	WFE; Later style square cab.
☐	061719-034	(E27N)	1/16	D	CHAD VALLEY		GB	52	WFE; Steerable; Clockwork; Rare.
☐	061719-035	(DDN) MAJOR	1/16	D	CHAD VALLEY		GB	54	WFE; Steerable; Hood side panels lift; Excellent detail.
☐	061719-036	DEXTA	1/16	D	CHAD VALLEY		GB	55	WFE.
☐	061719-037	MAJOR	1/43	D	CLIFFORD	CS-1	HK	68	WFE; Replica of Taisiya Fordson (061719-040).
☐	061719-038	MAJOR	1/20	D	CLIFFORD		HK	68	WFE; Steerable; Very good detail.
☐	061719-039	MAJOR	1/20	D	M.W.	5101	HK	62	WFE; Steerable; "Empire made" molded on fender.
☐	061719-040	MAJOR	1/43	D	TAISEIYA	501	J	65	WFE; With or without four-wheel wagon.
☐	061719-041	POWER MAJOR	1/62	D	LESNEY	72A	GB	59	WFE.
☐	061719-042	SUPER MAJOR	1/42	D	LESNEY	K-11	GB	63	WFE; With tandem trailer.
☐	061719-043A	POWER MAJOR	1/43	D	CORGI	55	GB	61	WFE; Headlights on sides of hood.
☐	061719-043B	POWER MAJOR	1/43	D	CORGI	54	GB	61	WFE; Headlights on sides of hood; With halftracks.
☐	061719-044A	POWER MAJOR	1/43	D	CORGI	60	GB	62	WFE; Headlights in grill.
☐	061719-044B	POWER MAJOR	1/43	D	CORGI	54	GB	62	WFE; Headlights in grill; With halftracks.
☐	061719-045	MAJOR	1/41	P	POLITOYS	65	I	63	WFE; Red and gray; With four-wheel wagon.
☐	061719-046	SUPER MAJOR	1/43	P	MINALUXE		F	67	WFE; Blue and silver.
☐	061719-047	SUPER MAJOR	1/43	P	TOMTE		N	75	WFE; Molded in driver; With four-wheel wagon.
☐	061719-048	(DEXTA)	1/25	D	CRESCENT	1805	GB	57	WFE; Available also with a four-wheel wagon; The color is orange in the farm set.
☐	061719-049	(DEXTA)	5-1/4"	P	LUCKY	184	HK	??	WFE; Gyro motor.
☐	061719-050	(DEXTA)	5-1/4"	P	MIC	D-6	HK	??	WFE; Battery operated; Driver.
☐	061719-051	N STADARD	1/32	WM KIT	BROWN'S MODELS		GB	76	WFE; Model of '44 English Fordson.
☐	061719-052	(DEXTA)	1/24	P	TOMTE		N	77	WFE; Variety of colors; Molded in driver.
☐	061719-053	F/LOADER	1/16	CI	ARCADE		USA	??	WFE; Rear crank operated front loader; Rare.
☐	061719-054	(F)	8"	T	BING		D	20	WFE; Clockwork.
☐	061719-055	(F)/HALF TRACKS	8"	T	BING		D	20	WFE; Clockwork; With halftracks.
☐	061719-056	(F)	5-1/2"	CI	A.C. WILLIAMS		USA	??	WFE; "Steel" wheels; Front crank; "Fordson" name cast along sides of radiator.
☐	061719-057	(F)	6"	CI	JOHN WRIGHT	20-401	USA	79	WFE; Reproduction of above.
☐	061719-058	(F)	1/20	CA KIT	ROY LEE BAKER	NONE	USA	81	WFE.
☐	061719-059	(F)	1/87	P KIT	THE WHEEL WORKS	WW 115	USA	80	WFE.
☐	061719-060A	E27N	1/32	WM KIT	BROWN'S MODELS		GB	80	WFE; Represents a '40's Fordson; "Steel" wheels.

	CODE	MODEL	SCALE	MATER.	MANU.	STOCK #	CTRY.	YEAR	REMARKS
☐	061719-060B	E27N	1/32	WM KIT	BROWN'S MODELS		GB	81	Same except on rubber; Also available as a "road" tractor with headlights, license plate, etc. for road work.
☐	061719-061A	E27N (HIGH MAJOR)	1/12	CUS.	KARSLAKE	NONE	GB	80	WFE; This model was used for roadwork; Very limited production.
☐	061719-061B	E27N (AGRICULTURAL)	1/12	CUS.	KARSLAKE	NONE	GB	82	WFE; Limited production; "Steel" wheels.
☐	061719-062	STANDARD	7-1/2"	T	METTOY	3262	GB	??	WFE; Some variations have clockwork motor while others have plastic engine.
☐	061719-063	MAJOR		P	BLUE BOX	#38 77385	HK	??	WFE; With four-bottom plow or loader; Yellow.
☐	061719-064	(E27N)	1/250	D	TRAFFICAST		GB	47	WFE; Used on architectural model of proposed new Fordson factory in Basildon, England to illustrated completed tractors in parking lot.
☐	061719-065	(MAJOR)	1/500	D	TRAFFICAST		GB	52	WFE; Used as an architectural model for British Ford Motor Company.
☐	061719-066	ALL AROUND	1/16	CA	SCALE MODELS	NO. 4	USA	80	Row crop; "Steel" wheels; Represents a 1936 Fordson; No. 4 in the JLE Collector Series; 3000 made.
☐	061719-067A	(F)	1/32	WM KIT	BROWN'S MODELS		GB	80	WFE; Represents a "water washer" air cleaner 1934 Fordson.
☐	061719-067B	(F)	1/32	WM KIT	BROWN'S MODELS		GB	80	WFE; Represents a 1944 model without "water washer" air cleaner.
☐	061719-068	(F)	1/16	CA	PIONEER TRACTOR WORKS		USA	80	WFE; American or English color versions; Fenders; Driver; (See also 060414-021
☐	061719-069	(F)/HALF TRACKS	6-7/8"	D	?		USA	??	WFE; "H.P.S. CO. BUCYRUS, OHIO - RIGID RAIL" embossed on sides; Rare.
☐	061719-070	(F)	5-1/2"	T	SG		D	??	WFE; Clockwork; With wagon, disk and rake.
☐	061719-071	E27N	1/16	CUS.	MARBIL	NONE	GB	83	WFE; Rubber tires; See also County; Limited production.
☐	061719-072	(F)		D	??		GB	34	WFE; Copy of Dinky Fordson; Hood can be raised.
☐	061719-073	(F)	1/43	WM	SUN MOTOR CO.	306	GB	84	WFE; Copy of Dinky Fordson; Hood can be raised.
☐	061719-074A	N	1/64	PK	FROG	19 P.G.M.	GB	??	WFE; Army green.
☐	061719-074B	N CRAWLER	1/64	PK	FROG		GB	??	Crawler.
☐	061719-075A	(F) CRAWLER	1/40	T	MINIC TRIANG	11M	GB	??	Crawler with clockwork.
☐	061719-075B	(F) CRAWLER	1/40	T	MINIC TRIANG	11M/CF	GB	??	Military crawler.
☐	061719-076	(F)	1/40	T	MINIC TRIANG	83M	GB	??	Clockwork; Green or red.
☐	061719-077	(DEXTA)	1/45	P	GRISONI GIOCATTOLI		I	??	WFE; Red or blue; Also available with blade or front loader.
☐	061719-078	(STANDARD)	9"	T	METTOY		GB	??	WFE; With clockwork; Driver.
☐	061719-079	E27N	1/43	D	CHAD VALLEY	9502	GB	55	WFE; Clockwork; Red & yellow; Made under license by Rabro & Sturdy Products (Pty.) Ltd., Johannesburg, S.A. but has "Chad Valley, G.B." embossed on tractor.
☐	061719-080	(F)	1-3/4"	L	ECCLES BROTHERS	134	USA	74	WFE; Replica of Tootsietoy Fordson.
☐	061719-081	(F)	2-5/8"	L	ECCLES BROTHERS	135	USA	74	WFE; Replica of Tootsietoy Fordson.
☐	061719-082	(F)	5-1/4"	CI	KENTON		USA	??	WFE; "Steel" wheels; Cast in driver; Cast in "water washer" air cleaner back of engine.
☐	061719-083A	MAJOR	1/32	WM KIT	SCALEDOWN	T-10	GB	83	WFE.
☐	061719-083B	MAJOR ROADLESS	1/32	WM KIT	SCALEDOWN	T-11	GB	83	WFE; Front wheel assist.
☐	061719-083C	MAJOR ROADLESS/ HALFTRACK	1/32	WM KIT	SCALEDOWN	T-12	GB	83	WFE; Half tracks.
☐	061719-083D	MAJOR DOE TRIPLE D	1/32	WM KIT	SCALEDOWN	T-13	GB	83	Two tractors with front axles removed and mounted in tandem being controlled by a single driver from the rear tractor.
☐	061719-084A	(N)	1/87		LINE BROS.		GB	??	WFE.
☐	061719-084B	COVENTRY CLIMAX	1/87		LINE BROS.		GB	??	Crawler conversion.
☐	061719-085	MAJOR	1/43	P	TIMPO		GB	??	WFE; "Chromed" engine.

061723-000 FOWLER

	CODE	MODEL	SCALE	MATER.	MANU.	STOCK #	CTRY.	YEAR	REMARKS
☐	061723-001	BIG LION	1/80	D	LESNEY	Y-9	GB	58	WFE; Steam traction Showman's engine.
☐	061723-002	SHOWMAN'S ENGINE	1/72	PK	HALES	K-302	GB	80	WFE; Steam traction engine.
☐	061723-003	PLOWING ENGINES	1/72	WM KIT	W & K		GB	80	WFE; Set of two plowing engines and a plow.
☐	061723-004	CLASS 27 PLOWING ENGINE	1/72	PK	KEIL KRAFT	K-304	GB	84	WFE; Steam traction plowing engine.
☐	061723-005	??	1/80	WM KIT	ANBMCS		GB	70	WFE; Copy of Lesney; Steerable.
☐	061723-006	ROAD TRACTION ENGINE	1/43	WM KIT	PHANTOM MODELS	A-23	F	84	WFE; Represents an antique steam traction engine for road use.

061809-000 FRICK

	CODE	MODEL	SCALE	MATER.	MANU.	STOCK #	CTRY.YEAR	REMARKS
☐	061809-001	16 HP	1/25	CA	BRUBAKER; WHITE; PETERSON; IRWIN	NONE	USA 64	WFE; Steam traction engine.
☐	061809-002	50 HP	1/93	CUS.	PHANTOM	A07F ?		

062214-000 FUNKSTORGRAD N
	CODE	MODEL	SCALE	MATER.	MANU.	STOCK #	CTRY.YEAR	REMARKS
☐	062214-001	10 KM CRAWLER	10''	TIN	M/S	3005	USSR ??	Crawler.

070118-000 GARNETT
	CODE	MODEL	SCALE	MATER.	MANU.	STOCK #	CTRY.YEAR	REMARKS
☐	070118-001	PENDLLE PRINCESS	1/16	PK	BANDAI	8025	J ??	WFE; Represents a 1919 Showman's engine.

070508-000 GEHL
	CODE	MODEL	SCALE	MATER.	MANU.	STOCK #	CTRY.YEAR	REMARKS
☐	070508-001	4610	1/25	D	NZG	236	D 83	Skid steer loader.

071802-000 GRAHAM BRADLEY
	CODE	MODEL	SCALE	MATER.	MANU.	STOCK #	CTRY.YEAR	REMARKS
☐	071802-001	32 HP	4''	R	AUBURN RUBBER		USA 39	Row crop; Molded in driver.

072109-000 GUIDART
	CODE	MODEL	SCALE	MATER.	MANU.	STOCK #	CTRY.YEAR	REMARKS
☐	072109-001	??	1/25	P	CURSOR		D 60	WFE.

072112-000 GULDNER
	CODE	MODEL	SCALE	MATER.	MANU.	STOCK #	CTRY.YEAR	REMARKS
☐	072112-001	SPRINTER I	1/32	P	CURSOR		D 60	WFE; Red with gray wheels.
☐	072112-002	TOLEDO	1/20	P	CURSOR		D 60	WFE.

080114-000 HANOMAG   NOW A PA OF MASSEY-FERGUSON LTD.
	CODE	MODEL	SCALE	MATER.	MANU.	STOCK #	CTRY.YEAR	REMARKS
☐	080114-002	PERFECT 400	1/32	P	CURSOR	366	D 67	WFE; Blue and gray.
☐	080114-003	55 CV	1/60	D	SIKU	V165	D 60	WFE; Industrial with dump wagon; Blue; Rare.
☐	080114-004A	ROBUST 900	1/50	D	SIKU	V-287	D 73	WFE; Pale green.
☐	080114-004B	ROBUST 900	1/50	D	SIKU	V287	D 73	WFE; Blue; Promo with special box.
☐	080114-004C	ROBUST 900	1/50	D	SIKU	V287	D 76	WFE; Orange over green; Different hood.
☐	080114-005A	K12C	1/50	D	CURSOR	1269	D 71	Industrial crawler with loader.
☐	080114-005B	K 12D	1/50	D	CURSOR	1269	D 71	Industrial bulldozer.
☐	080114-005C	D 600	1/50	D	CURSOR	1269	D 71	Industrial bulldozer.
☐	080114-005D	D 600 D	1/50	D	CURSOR	1269	D 71	Industrial bulldozer.
☐	080114-006	(R 440)	1/87	P	EKO	2114	E 70	WFE; Copy of Augulplas; Red or yellow.
☐	080114-007	??	5-3/4''	T	MARUSAN TOYS	5189	J 62	WFE; Construction road tractor; Called a Hanomag but actually only a toy.
☐	080114-008A	B16	1/50	P	CURSOR	569	D 70	Industrial wheel loader.
☐	080114-008B	B16C	1/50	D	CURSOR	569	D 70	Industrial wheel loader.
☐	080114-008C	66C	1/50	D	CURSOR	569	D 70	Industrial wheel loader.
☐	080114-009	R16	1/43	P	??		D 55	WFE; Green.
☐	080114-010	R12	1/50		??		D 57	WFE; Green.
☐	080114-011	(ST 100)	1/40	D	MERCURY	64	I 50	WFE; Copy of Marklin; Clockwork; Similar to STS 100 but not exactly a Hanomag.
☐	080114-012A	R45	1/87	PK	PREISER	600	D 84	WFE; Cab; Circus tractor.
☐	080114-012B	R45	1/87	PK	PREISER	912	D 84	WFE; With two trailers.
☐	080114-013A	ST 100	1/87	P	WIKING	502	D 82	WFE; With low bed trailer; Blue.
☐	080114-013B	ST 100	1/87	P	WIKING	3500	D 83	WFE; Road tractor with two trailers; Brown.
☐	080114-014	ST 100	1/43	WM KIT	STEAM & TRUCK	ST 05	D 83	WFE.
☐	080114-015A	SS 100	1/43	D	MARKLIN		D 38	WFE.
☐	080114-015B	SS 100	1/43	D	MARKLIN	8021/81	D 38	WFE; Military model.
☐	080114-016	??	1/25	P	WIKING		D 52	WFE; Very rare.
☐	080114-017	??	1/87	P	WIKING	T 38	D 52	WFE; Dark green; Rare.
☐	080114-018	ROBUST	1/30	P	??		H ??	WFE; With or without Robust name.
☐	080114-019	K 65E	1/60	P	SIKU	V99	D 60	Crawler with angle dozer; Red; Rare.
☐	080114-020	66C	1/80	BRONZE	PADRINI	NONE	I 83	Industrial wheel loader; Promo.
☐	080114-021	ROBUST	1/30	P	??		H ??	WFE; With or without Robust name.
☐	080114-022	R40 1942	1/43	WM KIT	STEAM & TRUCK		D 84	WFE.

080115-000 HANOMAG-BARREIROS   SEE ALSO HANOMAG
	CODE	MODEL	SCALE	MATER.	MANU.	STOCK #	CTRY.YEAR	REMARKS
☐	080115-001	R 440	1/87	P	AUGUPLAS-MINICARS	98	E 64	WFE; Red with blue wheels.

080519-000 HESSTON   NOW A PART OF FIAT CORP.
	CODE	MODEL	SCALE	MATER.	MANU.	STOCK #	CTRY.YEAR	REMARKS
☐	080519-001	880DT	1/32	D & P	BRITAINS LTD.	9528	GB 81	WFE; Identical to Fiat 880DT (060901-010) except having Hesston decals and colored environmental red; This model was used as a promotional for the introduction of the Hesston tractors in the United States in 1981.
☐	080519-002A	980	1/16	D & P	SCALE MODELS	980	USA 82	WFE; Cab.
☐	080519-002B	980DT	1/16	D & P	SCALE MODELS	980DT	USA 82	WFE; Front-wheel-assist; Cab.
☐	080519-002C	1380	1/16	D & P	SCALE MODELS	1380	USA 82	WFE; Duals.
☐	080519-002D	980DT COMM.	1/16	D & P	SCALE MODELS	980DT	USA 82	WFE; Same as 080519-002B but on a walnut plaque.
☐	080519-002E	1380 COMM.	1/16	D & P	SCALE MODELS	1380	USA 82	WFE; Same as 080519-002C but on a walnut plaque.
☐	080519-002F	980	1/16	D & P	SCALE MODELS	980	USA 84	WFE; Same as 080519-002A except no diagonal side trim.
☐	080519-002G	980DT	1/16	D & P	SCALE MODELS	980DT	USA 83	WFE; Same as 080519-002B except no diagonal side trim.
☐	080519-002H	1380	1/16	D & P	SCALE MODELS	1380	USA 83	WFE; Same as 080509-002C except no diagonal side trim.
☐	080519-003A	980DT	1/64	D & P	MINI-TOYS	301	USA 84	WFE; Front-wheel-assist; Cab.

CODE	MODEL	SCALE	MATER.	MANU.	STOCK #	CTRY.	YEAR	REMARKS
☐ 080519-003B	980	1/64	D & P	MINI-TOYS	302	USA	84	WFE; Cab.
☐ 080519-003C	1180DT	1/64	D & P	MINI-TOYS	303	USA	84	WFE; Front-wheel-assist; Cab.
☐ 080519-003D	1180	1/64	D & P	MINI-TOYS	304	USA	84	WFE; Cab; Duals.
☐ 080519-003E	1180DT	1/64	D & P	MINI-TOYS	305	USA	84	WFE; Front-wheel-assist; Cab; Duals.
☐ 080519-003F	1180	1/64	D & P	MINI-TOYS	306	USA	84	WFE; Cab; Duals.
☐ 080519-003G	1180 TURBO	1/64	D & P	MINI-TOYS	313	USA	84	WFE; Cab; Duals; Air cleaner; Special Edition; 5000 made.
☐ 080519-003H	100-90 DT	1/64	D & P	MINI-TOYS	307	USA	84	WFE; Front-wheel-assist; Cab.
☐ 080519-003I	100-90	1/64	D & P	MINI-TOYS	308	USA	84	WFE; Cab.
☐ 080519-003J	130 90 DT	1/64	D & P	MINI-TOYS	309	USA	84	WFE; Front-wheel-assist; Cab.
☐ 080519-003K	130-90	1/64	D & P	MINI-TOYS	310	USA	84	WFE; Cab.
☐ 080519-003L	130-90 DT	1/64	D & P	MINI-TOYS	311	USA	84	WFE; Front-wheel-assist; Duals; Cab.
☐ 080519-003M	130 90	1/64	D & P	MINI-TOYS	312	USA	84	WFE; Duals; Cab.

080914-000 HINDUSTAN

CODE	MODEL	SCALE	MATER.	MANU.	STOCK #	CTRY.	YEAR	REMARKS
☐ 080914-001	??	1/20	D	MAXWELL		IN	80	WFE; Yellow over orange.

081512-000 HOLDER

CODE	MODEL	SCALE	MATER.	MANU.	STOCK #	CTRY.	YEAR	REMARKS
☐ 081512-001	CULTITRAC A55	1/30	D	CURSOR	1076	D	78	Four-wheel-drive; Articulated; Small tractor.
☐ 081512-002A	CULTITRAC A60	1/27.5	D	CURSOR	880	D	80	WFE; Green over black; Cab.
☐ 081512-002B	CULTITRAC A60	1/27.5	D	CURSOR	880	D	82	WFE; Industrial; Cab; Orange over black.
☐ 081512-003	C-500	1/27.5	D	CURSOR	484	D	84	WFE; Four-wheel-drive; Cab; Articulated; Orange.

081514-000 HONDA

CODE	MODEL	SCALE	MATER.	MANU.	STOCK #	CTRY.	YEAR	REMARKS
☐ 081514-001	F 190 CULT.	1/12	PK	SANWA	224 TKK N.15	J	60	Two-wheel garden tractor.

082102-000 HUBER

CODE	MODEL	SCALE	MATER.	MANU.	STOCK #	CTRY.	YEAR	REMARKS
☐ 082102-001	STEAM ROLLER	1/25	CI	HUBLEY		USA	29	Steam traction engine/roller wheels.
☐ 082102-002A	RETURN FLU	1/25	CA	BRUBAKER; WHITE; PETERSON; IRVIN	NONE	USA	64	Modified reproductin of Hubley Huber 082102-001; Steam traction engine; Irvin's model has name cast in.
☐ 082102-002B	STEAM ROLLER	1/25	CA	BRUBAKER; WHITE; PETERSON; IRVIN	NONE	USA	64	Reproduction of Hubley Huber 082102-002A.
☐ 082102-003	HUBER WARCO 10-D	1/40	D	JUE	HO-002/1	BR	??	Industrial road grader.
☐ 082102-004	STEAM ROLLER	3-1/4"	CI	HUBLEY		USA	29	Steam traction engine with roller.

082519-000 HYSTER

CODE	MODEL	SCALE	MATER.	MANU.	STOCK #	CTRY.	YEAR	REMARKS
☐ 082519-001	STRADDLE TRUCK	1/16"	D	DRUGE BROS.		USA	50	Industrial life truck for lumber, etc.
☐ 082519-002	STRADDLE TRUCK	1/32	D	DRUGE BROS.		USA	50	Industrial lift truck for lumber, etc.
☐ 082519-003	FORKLIFT		D	TEKNO	864	DEN	??	Industrial fork lift.
☐ 082519-004	FORKLIFT	1/43	D	CORGI	1113	GB	78	Industrial fork lift.

090803-000 IH

IHC — McCORMICK-DEERING — FARMALL

CODE	MODEL	SCALE	MATER.	MANU.	STOCK #	CTRY.	YEAR	REMARKS
☐ 090803-001A	(10-20)	1/16	CI	ARCADE	276-X	USA	25	WFE; Standard; Molded in pulley; Wheel variations include "steel" spoke type or solid disk with small solid rubber tires; Gray.
☐ 090803-001B	(10-20)	1/16	CI	ARCADE	276-0	USA	25	WFE; Same except with movable belt pulley; Plated driver; Rubber tires/wheels; Colors include red, gray or green.
☐ 090803-002A	FARMALL REGULAR	1/16	CI	ARCADE	279-X	USA	26	Row crop painted driver; Gray with red "steel" wheels.
☐ 090803-002B	FARMALL REGULAR	1/16	CI	ARCADE	279-O	USA	34	Row crop; Same except with belt pulley and painted driver; White rubber tires/wheels.
☐ 090803-003	TRAC-TRACTOR (TD-40)	1/16	CI	ARCADE	277	USA	36	Crawler with plated driver; Metal or rubber tracks.
☐ 090803-004	TRAC-TRACTOR (TD-18)	1/16	CI	ARCADE		USA	41	Crawler; Plated driver.
☐ 090803-005	M	4-1/4"	CI	ARCADE	7321	USA	42	Row crop; Cast in driver; Wood wheels.
☐ 090803-006	M	5-1/4"	CI	ARCADE	2329	USA	41	Row crop; Cast in driver; Rubber wheels/tires.
☐ 090803-007	M	1/16	CI	ARCADE	7070	USA	40	Row crop; Plated driver; Decal variations include gold lettering or white lettering with a black border; Rubber wheels/tires.
☐ 090803-008	A	1/12	CI	ARCADE	7050	USA	41	WFE; Plated driver; Offset "Culti-Vision" design; Gold lettering on decals.
☐ 090803-009	(10-20)	1/20	CI	KILGORE		USA	??	WFE; Standard style; Plated driver; Rare.
☐ 090803-010A	(10-20)	1/16	CA or K	ROBERT GRAY	NONE	USA	70	WFE; Reproduction of Arcade (10-20); Large belt pulley.
☐ 090803-010B	(10-20)	1/16	KOR	ROBERT GRAY	NONE	USA	79	WFE; 10th Anniversary model; Marked with gold lettering; Gold driver.
☐ 090803-011A	(10-20)	1/16	CA	OLD TIME TOYS	NONE	USA	69	WFE; Reproduction of Arcade (10-20).
☐ 090803-011B	(10-20)	1/16	CA	PIONEER TRACTOR WORKS	NONE	USA	81	WFE; Reproduction of Arcade (10-20).
☐ 090803-012A	FARMALL REGULAR	1/16	CA	OLD TIME TOYS	NONE	USA	69	Row crop; Reproduction of Arcade Farmall Regular (090803-002A).

	CODE	MODEL	SCALE	MATER.	MANU.	STOCK #	CTRY.	YEAR	REMARKS
☐	090803-012B	FARMALL REGULAR	1/16	CA	PIONEER TRACTOR WORKS	NONE	USA	81	Row crop; Reproduction of Arcade Farmall Regular.
☐	090803-013	TRAC-TRACTOR	1/24	WK	MOD-AC MFG.	146-1	USA	??	Crawler; Wood kit; World War II era.
☐	090803-014A	(M)	1/16	P	PRODUCT MINIATURE	2853	USA	47	Row crop; Spoke front wheels; "McCormick Farmall" decals.
☐	090803-014B	(M)	1/16	PK	PRODUCT MINIATURE	2853	USA	47	Row crop; Front wheel variations include spoke or non-spoke types; Decal variations include "McCormick Farmall" or "International Harvester Farmall".
☐	090803-014C	(M)/STARS	1/16	P	PRODUCT MINIATURE	2853	USA	51	Row crop; White with red wheels and blue stars; Special issue.
☐	090803-014D	(M)	1/16	P	PRODUCT MINIATURE	2853	USA	47	Row crop; Wheel and color variations exist; No decals on color variations.
☐	090803-014E	M	1/16	P	PRODUCT MINIATURE		USA	51	Row crop; Like 090803-014 except having special decals identifying it as the Australian version.
☐	090803-015A	SUPER C	1/16	P	LAKONE		USA	52	Row crop; Spring under seat.
☐	090803-015B	SUPER C	1/16	P	LAKONE-CLASSIC	NONE	USA	83	Row crop; Re-issued by Classic Farm Toys using original molds.
☐	090803-016A	200	1/16	P	LAKONE		USA	53	Row crop; Spring under seat.
☐	090803-016B	230	1/16	P	LAKONE		USA	54	Row crop; White painted grill.
☐	090803-016C	200	1/16	P	LAKONE-CLASSIC	NONE	USA	83	Row crop; Re-issued by Classic Farm Toys using original molds.
☐	090803-016D	230	1/16	P	LAKONE-CLASSIC	NONE	USA	83	Row crop; Re-issued by Classic Farm Toys using original molds.
☐	090803-017	UD-24 POWER UNIT	1/16	P	PRODUCT MINIATURE		USA	50	Stationary power unit; Cigarette dispenser.
☐	090803-018A	TD-24 CRAWLER	1/16	P	PRODUCT MINIATURE		USA	49	Crawler; Decal variations include "International Diesel" and "International", The real TD-24 was introduced in 1947.
☐	090803-018B	TD-24 CRAWLER	1/16	P	PRODUCT MINIATURE		USA	49	Crawler/blade, Remote electric motor and lights.
☐	090803-018C	TD-24 CRAWLER	1/16	P	PRODUCT MINIATURE		USA	49	Crawler/blade; No motor and lights.
☐	090803-019	(M) DIESEL FIXALL	1/12	PK	MARX		USA	54	Row crop; Has set of tools and driver. Color variations include red, orange and gray.
☐	090803-020	H	1/20	D	HUBLEY		USA	52	Row crop; No driver.
☐	090803-021	(M)	1/12	D	HUBLEY		USA	52	Row crop; Steerable. Some had a mounted front loader or a mounted cultivator.
☐	090803-022	(M)	4"	R	AUBURN RUBBER		USA	??	Row crop; Molded in driver; Color variations.
☐	090803-023	(M)	6-1/2"	D	SLIK TOYS	8924	USA	??	Row crop.
☐	090803-024	(M)	6-1/2"	D	LINCOLN SPECIALITIES	9824	CDN	??	Row crop; Similar to Slik (M) (090803-023).
☐	090803-025	(M)	1/16	CA	ERTL		USA	45	Row crop; Cast in driver; Solid rubber, red center, wheels/tires.
☐	090803-026	(M)	1/16	CA	ERTL		USA	50	Row crop; Yellow wheels; No driver.
☐	090803-027A	CUB CADET 122	1/16	D	ERTL	432	USA	66	Lawn and garden tractor.
☐	090803-027B	CUB CADET 122	1/16	D	ERTL	433	USA	66	Lawn and garden tractor/blade.
☐	090803-028A	CUB CADET 125	1/16	D	ERTL	432	USA	68	Lawn and garden tractor.
☐	090803-028B	CUB CADET 125	1/16	D	ERTL	433	USA	68	Lawn and garden tractor/blade.
☐	090803-029A	CUB CADET 126	1/16	D	ERTL	432	USA	70	Lawn and garden tractor.
☐	090803-029B	CUB CADET 126	1/16	D	ERTL	433	USA	70	Lawn and garden tractor/blade.
☐	090803-030A	CUB CADET 129	1/16	D	ERTL	432	USA	72	Lawn and garden tractor.
☐	090803-030B	CUB CADET 129	1/16	D	ERTL	473	USA	72	Lawn and garden tractor/blade.
☐	090803-030C	CUB CADET 129	1/16	D	ERTL	474	USA	72	Lawn and garden tractor/blade and trailer.
☐	090803-030D	CUB CADET 129	1/16	D	ERTL	473	USA	76	Lawn and garden tractor; "Spirit of 76" Special Issue; Red, white and blue.
☐	090803-031A	CUB CADET 1650	1/16	D	ERTL	435	USA	76	Lawn and garden tractor/covered engine.
☐	090803-031B	CUB CADET 1650	1/16	D	ERTL	436	USA	76	Lawn and garden tractor/blade and trailer; Covered engine.
☐	090803-032A	240	1/16	D	ERTL		USA	59	WFE; Utility; With fast-hitch.
☐	090803-032B	340	1/16	D	ERTL		USA	59	WFE; Utility; With fast-hitch.
☐	090803-033	340 INDUSTRIAL	1/16	D	ERTL		USA	59	WFE; Utility; With fast-hitch; Decal grill; Yellow.
☐	090803-034A	400	1/16	D	ERTL		USA	54	Row crop; With fast-hitch; Split rear rims; Small rear tires.
☐	090803-034B	400	1/16	D	ERTL		USA	55	Row crop; With fast-hitch.
☐	090803-035	404	1/16	D	ERTL		USA	61	WFE; Utility; Three-point-hitch; Decal variations include "International" and headlights in the grill decal.
☐	090803-036A	404	1/16	D	ERTL	437	USA	64	Row crop; Red metal or plastic wheels; Decal variations include "Farmall".
☐	090803-036B	404	1/16	D	ERTL	437	USA	67	Row crop; White plastic wheels.
☐	090803-037	450	1/16	D	ERTL		USA	56	Row crop; With fast-hitch.

CODE	MODEL	SCALE	MATER.	MANU.	STOCK #	CTRY.	YEAR	REMARKS
☐ 090803-038A	460	1/16	D	ERTL		USA	58	Row crop; With fast-hitch or without fast-hitch; With or without belt pulley; Front and rear wheel variations.
☐ 090803-038B	560	1/16	D	ERTL	408	USA	57	Row crop; With fast-hitch; Early version has 450 type front wheels.
☐ 090803-038C	560	1/16	D	ERTL	408	USA	64	Row crop; No fast-hitch; Front wheel variations; Either metal or plastic wheels.
☐ 090803-038D	560	1/16	D	ERTL	408	USA	67	Row crop; White wheels; No fast-hitch.
☐ 090803-038E	560	1/16	D	ERTL	460	USA	67	Row crop; Dual white, rear wheels; No fast-hitch.
☐ 090803-038F	560	1/16	D	ERTL	409	USA	68	Row crop; Cab; White wheels; No fast-hitch, pulley or exhaust; Two cab style variations.
☐ 090803-038G	(560)	1/16	D	ERTL	408	USA	72	Row crop; No. 560 decals, exhaust, fast-hitch or pulley; Plastic wheels.
☐ 090803-038H	560 TOY FARMER	1/16	D	ERTL		USA	78	WFE; Dated 11/1/78; 500 made for the National Farm Toy Show.
☐ 090803-038I	560 TOY FARMER	1/16	D	ERTL		USA	79	WFE; Duals; Dated 3/1/79; 500 made.
☐ 090803-039A	544	1/16	D	ERTL	414	USA	69	WFE; Red plastic wheels.
☐ 090803-039B	(544)	1/16	D	ERTL	415	USA	69	Row crop; No. 544 decals.
☐ 090803-039C	(544)	1/16	D	ERTL	417	USA	69	Row crop; Dual white, rear wheels.
☐ 090803-039D	(544)	1/16	D	ERTL	418	USA	69	Row crop/loader; White wheels.
☐ 090803-039E	(2644) INDUSTRIAL	1/16	D	ERTL	421	USA	70	Row crop/loader; Yellow.
☐ 090803-039F	(2644) INDUSTRIAL	1/16	D	ERTL	416	USA	69	WFE; Yellow.
☐ 090803-039G	(544)	1/16	D	ERTL		USA	81	Row crop; White plastic wheels/tires.
☐ 090803-040A	FLYING FARMALL	1/16	D	ERTL	5855	USA	74	Row crop; Modified 560 Super Rod; Large front wheels; Red.
☐ 090803-040B	FLYING FARMALL	1/16	D	ERTL	5855	USA	74	Row crop; Same except small front wheels; Red.
☐ 090803-040C	FLYING FARMALL	1/16	D	ERTL	5855	USA	74	Row crop; Same except large rear wheels; Maroon.
☐ 090803-041A	FARMALL 656	1/32	D	ERTL	40	USA	67	Row crop; Non-steering.
☐ 090803-041B	INTERNATIONAL 656	1/32	D	ERTL	40	USA	68	Row crop; Non-steering; Some did not have model designation on decals.
☐ 090803-042A	(666)	1/32	D	ERTL	405	USA	74	Row crop; Non-steering.
☐ 090803-042B	(666)	1/32	D	ERTL	405	USA	76	WFE; Non-steering.
☐ 090803-043A	806	1/16	D	ERTL	435	USA	64	Row crop; Metal or plastic rear wheels; Round fenders.
☐ 090803-043B	806	1/16	D	ERTL	435	USA	65	Row crop; Plastic rear wheels; Flat fenders.
☐ 090803-043C	806	1/16	D	ERTL	435	USA	66	Row crop; Plastic rear wheels; Flat fenders.
☐ 090803-044A	856	1/16	D	ERTL	419	USA	68	WFE.
☐ 090803-044B	1026 HYDRO	1/16	D	ERTL	401	USA	71	WFE.
☐ 090803-044C	1256 TURBO	1/16	D	ERTL	420	USA	68	WFE; Cab; Dual rear wheels; Front suitcase style weights.
☐ 090803-044D	1456 TURBO	1/16	D	ERTL	420	USA	71	WFE; Cab; Dual rear wheels; Front suitcase style weights.
☐ 090803-045A	966 HYDRO	1/16	D	ERTL	401	USA	72	WFE.
☐ 090803-045B	966	1/16	D	ERTL	403	USA	75	WFE; Cab; Dual rear wheels; Very few made.
☐ 090803-045C	1066 TURBO	1/16	D	ERTL	402	USA	72	WFE; ROPS.
☐ 090803-045D	1066 TURBO	1/16	D	ERTL	411	USA	75	WFE; Cab.
☐ 090803-045E	1466	1/16	D	ERTL	403	USA	72	WFE; Cab; Dual rear wheels.
☐ 090803-045F	1466	1/16	D	ERTL	403	USA	81	WFE; ROPS; Dual rear wheels.
☐ 090803-045G	966 HYDRO	1/16	D	ERTL	401	USA	71	WFE; Same as 090803-045H except having white front wheels.
☐ 090803-046	1206 TURBO	1/16	D	ERTL	436	USA	66	WFE; White wheels and trim.
☐ 090803-047	1466	1/25	PK	ERTL	8003	USA	74	WFE; Cab.
☐ 090803-048A	1466	1/64	D	ERTL	1355	USA	74	WFE; Without rivet through cab.
☐ 090803-048B	1466	1/64	D	ERTL	1355	USA	76	WFE; Cab; With rivet through cab.
☐ 090803-049A	886	1/16	D	ERTL	461	USA	76	WFE; Four post type ROPS; Front wheel variations.
☐ 090803-049B	1086	1/16	D	ERTL	462	USA	76	WFE; Cab; Front wheel variations.
☐ 090803-049C	1586	1/16	D	ERTL	463	USA	76	WFE; Cab; Dual rear wheels; Front wheel variations.
☐ 090803-050	TD 25	1/25	D	ERTL	427	USA	61	Industrial crawler/blade; Lights on top of radiator; Yellow with either black or red "International" decals.
☐ 090803-051	TD-25	1/25	D	ERTL	452	USA	71	Industrial crawler/blade; Lights on side of radiator; Yellow.
☐ 090803-052A	2504	1/16	D	ERTL	434	USA	63	WFE; Industrial utility; Three-point-hitch; Decal variations include early one with white side panel.
☐ 090803-052B	2504	1/16	D	ERTL	434	USA	64	WFE; Industrial utility without three-point-hitch.
☐ 090803-053A	3414	1/16	D	ERTL	428	USA	67	WFE; Industrial backhoe/loader.
☐ 090803-053B	3444	1/16	D	ERTL	428	USA	69	WFE; Industrial backhoe/loader.
☐ 090803-054	(3400)	1/16	D	ERTL	472	USA	75	WFE; Industrial backhoe/loader.

CODE	MODEL	SCALE	MATER.	MANU.	STOCK #	CTRY.	YEAR	REMARKS
090803-055A	HOUGH PAYLOADER	1/25	D	ERTL	426	USA	68	Industrial four-wheel-loader; Yellow.
090803-055B	HOUGH PAYLOADER	1/25	D	ERTL	426	USA	81	Industrial four-wheel-loader; Light brown.
090803-056	PAYHAULER	1/16	D	ERTL	419	USA	60	Industrial dump truck; Steerable; Dual rear wheels; Some have rubber exhaust stack.
090803-057A	(180) PAYHAULER	1/32	D	ERTL	425	USA	68	Industrial dump truck; Hydraulic dump; Dual wheels both front and rear. Yellow.
090803-057B	(180) PAYHAULER	1/32	D	ERTL	425	USA	68	Industrial dump truck; Same except with white cab.
090803-057C	(180) PAYHAULER	1/32	D	ERTL	425	USA	82	Industrial dump truck; Same except light brown.
090803-058	350 PAYHAULER	1/25	D	ERTL	8013	USA	74	Industrial dump truck; Yellow; Duals front and rear; Excellent detail.
090803-059	CUB	1/16	PK	AFINSON		USA	50	WFE; Offset "Culti-vision" style
090803-060	CUB	1/16	PK	ATMA		BR	50	WFE; Offset "Culti-vision" style; Raised letters in place of decals..
090803-061	CUB	1/16	PK	DESIGN FABRICATORS		USA	50	WFE; Offset "Culti-vision" style.
090803-062	CUB	1/16	PK	SAUNDERS-SWADER		USA	50	WFE; Offset "Culti-vision" style; Marketed by Reuhl Products.
090803-063	(M)	4"	T	MARX		J	??	Row crop; Plastic driver; Gyro motor; Wheel variations; With or without "Farmall" decals.
090803-064	(M) FUN HO	11-1/2"	D	FUN-HO	531	NZ	??	WFE; With loader; Green and silver.
090803-065A	(240)	1/20	T & P	UNIVERSAL-CORDEG	5101	HK	??	WFE; Utility; Red with white wheels.
090803-065B	(240)	1/20	D & P	UNIVERSAL-CORDEG	5102	HK	??	WFE; Utility; With driver; Red with yellow wheels.
090803-066A	B-250	1/37	D	LESNEY	K-4	GB	60	WFE; Utility; Red with green wheels.
090803-066B	B-250	1/37	D	LESNEY	K-4	GB	61	WFE; Same except red wheels and different rear hitch.
090803-067	(414)	1/32	D	TEKNO	465	DK	66	WFE; Utility; Headlights; Decal marked "McCormick International".
090803-068	(574)	1/32	D	TEKNO	466	DK	71	WFE; Utility/ROPS; Set no. 467 includes front blade and rotary brush.
090803-069	560	1/16	T	ALPS TOYS	5855	J	73	WFE; Smoking tractor; Tan and red; Does not look like the I.H. 560 tractor.
090803-070A	TD-25	1/80	D	MINI-DINKY	94	HK	69	Industrial crawler/blade.
090803-070B	TD-25	1/80	D	MINI-DINKY		HK	69	Industrial crawler.
090803-070C	TD-25	1/80	D	MINI-DINKY	85	HK	69	Industrial crawler/Drott loader.
090803-071	HOUGH PAYLOADER	1/80	D	MINI-DINKY		HK	69	Industrial four-wheel-loader; White.
090803-072	TD-25	1/80	D	MERCURY LITLE TOY	514/A	USA	??	Industrial crawler.
090803-073	HYDRAULIC EXCAVATOR	1/55	D	SOLIDO	365	F	75	Industrial excavator; Individual metal crawler track sections.
090803-074A	844	1/43	D	ELIGOR	3003	F	76	WFE; Front-wheel-assist; Cab; Weights.
090803-074B	844	1/43	D	ELIGOR		F	79	WFE; Same except marketed under the name "Diano".
090803-074C	844 or 953	1/43	D	ELIGOR		F	83	WFE; Same except marketed under the name "Confradis".
090803-075A	FARMALL F-30	8"	K	ROBERT GRAY	NONE	USA	76	Row crop; Represents a 1930's model.
090803-075B	FARMALL F-30	8"	K	OLD TIME COLLECTIBLES		USA	80	Row crop; Same except different manufacturer.
090803-076A	McCORMICK W-9	8-1/2"	K	ROBERT GRAY	NONE	USA	77	WFE; Standard style, 1939 model.
090803-076B	McCORMICK W-9	8-1/2"	K	OLD TIME COLLECTIBLES	NONE	USA	81	WFE; Same except different manufacturer.
090803-076C	McCORMICK W-9	8-1/2"	K	OLD TIME COLLECTIBLES	NONE	USA	83	WFE; Same except decalled and with rubber tires.
090803-077	FARMALL	6-1/2"	T	MARX		J	??	WFE; Front loader raises and lowers as tractor pushed along.
090803-078	560 PAYLOADER	4-3/4"	D	TOMY	4519	J	76	Industrial four-wheel-articulated loader.
090803-079	1086	1/64	D	ERTL	1620	USA	78	WFE; Cab.
090803-080	CUB	1/12	CA	?		USA	??	WFE; Offset "Culti-vision" style; May be one of a kind.
090803-081	FARMALL (C)	8"	P	THOMAS TOYS			50	Row crop.
090803-082A	3588 (2 + 2)	1/16	D	ERTL	464	USA	79	Four-wheel-drive; Articulated; "FIRST EDITION"; engraved on lower left front frame. No 2 + 2 decal. Only one available with display stand for each participating I.H. dealer; 3200 made.
090803-082B	3588 2 + 2	1/16	D	ERTL	464	USA	79	Same except "FIRST EDITION" ground out.
090803-082C	3588 2 + 2	1/16	D	ERTL	464	USA	69	Four-wheel-drive; Articulated; Regular issue.
090803-082D	6388 2 + 2	1/16	D	ERTL	464	USA	83	Four-wheel-drive; Articulated; Gray wheels with black inserts; Red cab.

	CODE	MODEL	SCALE	MATER.	MANU.	STOCK #	CTRY.	YEAR	REMARKS
☐	090803-083	(FARMALL M)	1/16	CA	SCALE MODELS	NO. 2	USA	79	Row crop; "Steel" wheels; COLLECTOR SERIES; 3000 made; Represents a 1939 model; First few hundred were issues with decals; No. 2 in the JLE Collector Series.
☐	090803-084A	IHC GAS ENGINE	1/16	CA	ROBERT GRAY	NONE	USA	81	Antique style portable gasoline engine; No wheels.
☐	090803-084B	IHC GAS ENGINE	1/16	CA	ROBERT GRAY	NONE	USA	81	Same except with wheels and hitch.
☐	090803-085A	IHC TYPE A	1/64	D & P	ERTL	1750	USA	80	WFE; Antique Titan style; Gray flywheel and steering wheel.
☐	090803-085B	IHC TYPE A	1/64	D & P	ERTL	1750	USA	80	WFE; Same except green flywheel and steering wheel.
☐	090803-086	TITAN 10-20	1/64	D	ERTL	1748	USA	80	WFE; Antique Mogul style tractor.
☐	090803-087	McCORMICK 10-20	1/64	D	ERTL	1749	USA	80	WFE; Represents a 1921 standard style tractor.
☐	090803-088	McCORMICK FARMALL	1/64	D	ERTL	1751	USA	80	Row crop; Represents a 1924 style tractor.
☐	090803-089	FARMALL (H)	1/64	D	ERTL	1747	USA	80	Row crop; Represents a 1939 tractor.
☐	090803-090	TD-18	1/16	CA	TOY & TACKLE		USA	??	Crawler; Arcade molds used.
☐	090803-091A	CUB CADET 682	1/16	D	ERTL	459	USA	80	Lawn and garden tractor; Red; "FIRST EDITION".
☐	090803-091B	CUB CADET 682	1/16	D	ERTL	459	USA	80	Same except regular issue; White or cream color wheels.
☐	090803-091C	CUB CADET	1/16	D	ERTL	465	USA	80	Lawn and garden tractor; Same except with blade and trailer.
☐	090803-091D	CUB CADET (682)	1/16	D	ERTL	499	USA	82	Lawn and garden tractor; Same except Cub Cadet Corporation version; Yellow.
☐	090803-091E	CUB CADET (682)	1/16	D	ERTL	699	USA	82	Lawn and garden tractor; Same except "LIMITED EDITION" CCC version; Yellow.
☐	090803-092	B-275 MAHINDRA	1/20	D	MAXWELL		IN	81	WFE; Made in India under license; Prominant exhaust and air cleaner.
☐	090803-093	423	1/28	P	HAUSSER	4450	D	80	WFE; Utility.
☐	090803-094	984	1/32	D	ERTL	1638	USA	81	WFE; Front-wheel-assist; Cab.
☐	090803-095	624 DIESEL	11"	P	MONT BLANC	659	F	65	WFE; Remote control.
☐	090803-096	844	1/87	PK	PREISER	950	D	79	WFE; With Landsberg plow.
☐	090803-097	(946)	1/32	D	LONE STAR	1702	GB	79	WFE; Color variations.
☐	090803-098	844-S	1/30	P	BRUDER	8305	D	81	WFE; With side dump trailer; Also in no. 103099 Farm Set.
☐	090803-099	1055	1/41	D & P	POLISTIL	CE-113	I	80	WFE; Front-wheel-assist; Cab.
☐	090803-100A	5088	1/16	D	ERTL	487	USA	81	WFE; Cab; Dual gray rear wheels with black inserts; "FIRST EDITION KC 9-81" engraved on the frame.
☐	090803-100B	5088	1/16	D	ERTL	468	USA	81	WFE; Cab; Gray rear wheels with black inserts; "SPECIAL EDITION 12-81" engraved on frame.
☐	090803-100C	5088	1/16	D	ERTL	468	USA	81	WFE; Cab.
☐	090803-100D	5288	1/16	D	ERTL	487	USA	81	WFE; Cab; Duals.
☐	090803-100E	5288	1/16	D	ERTL	409TW	USA	84	WFE; Front-wheel-assist; "SPECIAL EDITION, MAY 1984".
☐	090803-100F	5488	1/16	D	ERTL	409PA	USA	84	WFE; Front-wheel-assist; "FIRST EDITION, MAY 1984".
☐	090803-100G	5488	1/16	D	ERTL	409DO	USA	84	WFE; Front-wheel-assist; Cab; Dual rear wheels.
☐	090803-101	412B	1/64	D	ERTL	1855	USA	81	Industrial scraper pan.
☐	090803-102A	270 BACKHOE-LOADER	1/64	D	ERTL	1853	USA	79	Industrial backhoe/loader; Yellow.
☐	090803-102B	270 BACKHOE-LOADER	1/64	D	ERTL	1853	USA	81	Industrial backhoe/loader; Light brown.
☐	090803-103A	350 PAYHAULER	1/80	D	ERTL	1852	USA	79	Industrial dump truck; Yellow.
☐	090803-103B	350 PAYHAULER	1/80	D	ERTL	1852	USA	81	Industrial dump truck; Light brown.
☐	090803-104A	TD-20E	1/64	D	ERTL	1851	USA	79	Industrial crawler/blade; Yellow.
☐	090803-104B	TD-20E	1/64	D	ERTL	1851	USA	81	Industrial crawler/blade; Light brown.
☐	090803-105A	560 PAYLOADER	1/80	D	ERTL	1850	USA	79	Industrial; Yellow.
☐	090803-105B	560 PAYLOADER	1/80	D	ERTL	1850	USA	81	Industrial; Light brown.
☐	090803-106A	640 EXCAVATOR	1/64	D	ERTL	1854	USA	80	Industrial; Yellow.
☐	090803-106B	640 EXCAVATOR	1/64	D	ERTL	1854	USA	80	Industrial; Light brown.
☐	090803-107	BACKHOE-LOADER	1/16	P	ERTL	3095	USA	84	Industrial; Crude; Yellow.
☐	090803-108	65 PAYLOADER	1/50	D	DIAPET	K-7	J	80	Industrial forklift; Orange.
☐	090803-109A	5088	1/64	D	ERTL	1797	USA	82	WFE; Cab; Black cab posts; Gray wheels.
☐	090803-109B	5088	1/64	D	ERTL	1797	USA	83	WFE; Cab; Red wheels.
☐	090803-109C	5088	1/64	D	ERTL	1797	USA	84	WFE; Cab; Gray wheels.
☐	090803-110	1086	1/16	P	ERTL	029	USA	81	WFE; Remote control.
☐	090803-111	FARMALL (C)	1/65	P	JOUEF		F	59	Row crop; No driver.
☐	090803-112A	FARMALL (C)	1/30	P	JOUEF	213	F	54	Row crop; Color variations.
☐	090803-112B	FARMALL (C)	1/30	P	JOUEF	226	F	54	Row crop; Clockwork; With trailer; Color variations.
☐	090803-113	FARMALL (C)	1/16	P	JOUEF	235	F	54	Row crop; Clockwork; Removable

	CODE	MODEL	SCALE	MATER.	MANU.	STOCK #	CTRY.YEAR	REMARKS
☐	090803-114A	A	1/32	WM KIT	SCALEDOWN	T-3	GB 82	driver; With "FARMALL" decals or raised letters.
☐	090803-114B	B	1/32	WM KIT	SCALEDOWN	T-4	GB 82	WFE; Offset "Culti-vision" style.
☐	090803-115	SUPER M-TA	1/16	CUS.	PETE FREIHEIT	NONE	USA 83	Row crop with single or dual front wheels.
☐	090803-116A	H	1/16	CUS.	PETE FREIHEIT	NONE	USA 82	Row crop; Excellent detail.
☐	090803-116B	SUPER H	1/16	CUS.	PETE FREIHEIT	NONE	USA82	Row crop; Excellent detail.
☐	090803-117	140	1/16	CUS.	PETE FREIHEIT	NONE	USA80	WFE; Offset "Culti-vision" style; Excellent detail.
☐	090803-118	300	1/16	CUS.	PETE FREIHEIT	NONE	USA80	Row crop; Excellent detail.
☐	090803-119	350	1/16	CUS.	PETE FREIHEIT	NONE	USA80	Row crop; Excellent detail.
☐	090803-120	230	1/16	CUS.	PETE FREIHEIT	NONE	USA80	Row crop; Excellent detail.
☐	090803-120	230	1/16	CUS.	PETE FREIHEIT	NONE	USA 80	Row crop; Excellent detail.
☐	090803-121	M	1/16	CA KIT	ROY LEE BAKER	NONE	USA 83	Row crop; With or without steerable front wheels.
☐	090803-122	400	1/16	CA KIT	ROY LEE BAKER	NONE	USA 83	Row crop; With or without steerable front wheels.
☐	090803-123	M	1/32	CA KIT	ROY LEE BAKER	NONE	USA 83	Row crop.
☐	090803-124	400	1/32	CA KIT	ROY LEE BAKER	NONE	USA 83	Row crop.
☐	090803-125A	2 + 2 3588	1/64	D	ERTL	1526	USA 82	Four-wheel-drive; Articulated; White cab.
☐	090803-125B	2 + 2 6388	1/64	D	ERTL	1526	USA 83	Four-wheel-drive; Articulated; Red cab.
☐	090803-126A	FARMALL F-12	1/16	CA	A T & T COLLECTABLES	NONE	USA 82	Row crop; "Steel" wheels; Gray.
☐	090803-126B	FARMALL F-14	1/16	CA	A T & T COLLECTABLES	NONE	USA 82	Row crop; "Steel" wheels; Red.
☐	090803-126C	F-14	1/16	CA	A T & T COLLECTABLES	NONE	USA 83	Row crop with a single front wheel; "Steel" wheels.
☐	090803-126D	F-14	1/16	CA	A T & T COLLECTABLES	NONE	USA 83	Row crop; Rubber tires; Red.
☐	090803-127A	A	1/16	CUS.	LYLE DINGMAN	NONE	USA 82	WFE; Offset "Culti-vision" style. Excellent detail.
☐	090803-127B	B	1/16	CUS.	LYLE DINGMAN	NONE	USA 82	Row crop; Also a BN version.
☐	090803-128A	8-16	1/16	CA	ALVIN EBERSOL	NONE	USA 82	WFE; Represents a 1917 model; Green with red wheels; 100 made.
☐	090803-128B	8-16	1/16	CA	ALVIN EBERSOL	NONE	USA 82	WFE; Same except gray with red wheels.
☐	090803-129A	8-16 MOGUL	1/16	CA	SCALE MODELS	NO. 2	USA 82	WFE; JLE Threshers Series; Green with red wheels.
☐	090803-129B	10-20 MOGUL	1/16	CA	SCALE MODELS		USA 83	WFE; JLE Specialities Series; Has fuel tank on top center of chasis; Green with red wheels; Has decals.
☐	090803-130	955	1/43	D	YAXON	030	I 82	WFE; Front-wheel-assist; Cab.
☐	090803-131	(10-20)	1/16	T	BING		GER 20	WFE; Clockwork; With half-tracks.
☐	090803-132	844	15"	P	?		??	
☐	090803-133	(5288)	50 ML	POR	PACESETTER	NO. 2	USA 83	WFE; Decanter; 3600 made; "LITTLE RED MACHINE".
☐	090803-134	(5288)	750 ML	POR	PACESETTER	NO. 2	USA 83	WFE; Decanter; 6000 made "BIG RED MACHINE".
☐	090803-135	1931 REGULAR	1/16	CA	SCALE MODELS	NO. 10	USA 83	Row crop; "Steel" wheels; 3000 made; No. 10 in the JLE Collector Series; Gray with red wheels.
☐	090803-136	(3088)	1/16	P	ERTL	415EO	USA 84	WFE; Crude.
☐	090803-137	(H)	1-1/2"	P	?		USA ??	Row crop; Blue; On a standup base.
☐	090803-138	(M)	1/16	P	KEMP PRODUCTS LTD.		CDN 50	Row crop; Similar to Product Miniature (M) 090803-014.
☐	090803-139	6388 2 + 2	1/40	P	ERTL	4791DO	USA 83	Four-wheel-drive; Articulated; Radio controlled.
☐	090803-140	D-430	1/32	P	G & S ERZEUGNIS		D 57	WFE.
☐	090803-141	1933 F-12	1/16	CA KIT	ROY LEE BAKER		USA 84	Row crop; "Steel" wheels.
☐	090803-142	4386 & 4786	1/16	CUS.	DAVID SHARP		USA 79	Four-wheel-drive; Cab; Duals.
☐	090803-143A	3180	1/43	D & P	NPS	3009	D-HK 84	WFE; Front-wheel-assist; Gyro motor; Red and white; Cab.
☐	090803-143B	3180	1/43	D & P	NPS	3009	D-HK 84	WFE; Front-wheel-assist; Gyro motor; Blue and white; Cab.
☐	090803-144	F-20	1/16	CA	SCALE MODELS	No. 2	USA 84	Row crop; "Steel" wheels; Red with red wheels; NO. 2 JLE Collector Series II; 5000 made.
☐	090803-145	2+2	1 Liter	PORC	PACESETTER	NO. 2	USA 84	Four-wheel-drive; Decanter; "BIG RED MACHINE".
☐	090803-146	2+2	200 ML	PORC	PACESETTER	NO. 2	USA 84	Four-wheel-drive; Decanter; "BIG RED MACHINE".
☐	090803-147	(666)	1/32	P	COFALU		F ??	WFE; Copy of Ertl (666) 090803-042B.
☐	090803-148	??	1/43	P	VINYL LINE	1421	D ??	WFE; Red or blue; With trailer.
☐	090803-149	4156	1/66	D	POLISTIL	RJ127	I ??	WFE; Yellow.
☐	090803-150	(McCORMICK)	1/20	TIN	SSS	S1207	J ??	Two different types.
☐	090803-151	FARMALL (C)	1/65	P	BONUX		F ??	Row crop; Called "Red" tractor; Similar to Jouef (C).
☐	090803-152	FARMALL (C)	1/65	P	?		F ??	Row crop; Key ring.

CODE	MODEL	SCALE	MATER.	MANU.	STOCK #	CTRY.	YEAR	REMARKS
☐ 090803-153	SUPER C	1/16	CUS.	PIONEER TRACTOR WORKS	NONE	USA	82	Row crop; Detailed version of 090803-015A.
☐ 090803-154A	FARMTOY 806	1/16	D	SCALE MODELS		USA	82	Row crop; Similar to Ertl Farmall 806; Tire, grill and decal variations.
☐ 090803-154B	FARMTOY 1206	1/16	D	SCALE MODELS		USA	82	Row crop; Similar to Ertl Farmall 1206 Turbo; Tire, grill and decal variations.
☐ 090803-155	F-30	1/10	WOOD	KRUSE	NONE	USA	81	Row crop; Rubber.
☐ 090803-156	10 HP	1/10	WOOD	KRUSE	NONE	USA	77	Gasoline engine.
☐ 090803-157	300	1/16	D	ERTL		USA	84	Row crop; Steerable; "7th Annual National Farm Toy Show;; tractor.
☐ 090803-158	(M)	1/16	P	??		GB	??	WFE; Similar to Product Miniature (M) except having a wide front axle. Has Doncaster (England) on decals.
☐ 090803-159	(10-20)	1/16	CA	SCALE MODELS	NO. 13	USA	84	WFE; "Steel wheels"; No. 13 in the JLE Collector Series; 5000 made.
☐ 090803-160	(H)	11-1/4"	WOOD	DEE BROTHERS	NONE	USA	46	Row crop; Steerable; Available with a four-wheel wagon, also 11-1/4" long; Tractor decal "FARMTOY"; Wagon decal "DEE BROS FARM TOYS".
☐ 090803-161	REGULAR	1/16	CA	A T & T COLLECTABLES	NONE	USA	83	Row crop; "Steel" wheels; Gray.
☐ 090803-162	F-20	1/16	CA	A T & T COLLECTABLES	NONE	USA	83	Row crop; Rubber tires; Red.
☐ 090803-163	F-30	1/16	CA	A T & T COLLECTABLES	NONE	USA	83	Row crop; Rubber tires; Red.
☐ 090803-164	M	1/12	WOOD	GUBBELS	NONE	USA	80	Row crop; "Steel" wheels.
☐ 090803-165	F 30	1/10	WOOD	KRUSE	NONE	USA	79	Row crop; "Steel" wheels.
☐ 090803-166	F 20	1/10	WOOD	KRUSE	NONE	USA	81	Row crop; "Steel" wheels.
☐ 090803-167	M-TA	1/5	CUS.	ROY LEE BAKER	NONE	USA	81	Row crop; Represents a 1954 model.
☐ 090803-168	F-12	1/5	CUS.	ROY LEE BAKER	NONE	USA	83	Row crop; "Steel" wheels; Represents a 1933 model.
☐ 090803-169A	7488 2+2	1/16	D	ERTL		USA	84	WFE; Four-wheel-drive; Articulated; Cab; Collectors LIMITED EDITION. With duals, plastic grill and gray exhaust.
☐ 090803-169B	7488 2+2	1/16	D	ERTL	467DO	USA	84	WFE; Single wheels, metal grill and black exhaust.
☐ 090803-170	SUPER M-TA	1/16	CUS.	DENNIS PARKER	NONE	USA	82	Row crop; Very good detail.
☐ 090803-171	(66 SERIES)	4"	P	??			??	WFE.
☐ 090803-172	BACKHOE	1/16	P	ERTL		USA	84	WFE; Industrial backhoe and loader; Crude.
☐ 090803-173	C	1/16	P	TUDOR ROSE		GB	??	Row crop; Variety of colors.
☐ 090803-174	(B-250)	12 cm	P	??			??	WFE.
☐ 090803-175	844	1/43	D	??			??	WFE.
☐ 090803-176	(M)	1/16	D	CARTER TRU-SCALE		USA	50	Row crop; Free turning front axle; Similar in appearance to Ertl (M) 090803-026.
☐ 090803-177	300	1/43	D	ERTL	2513	USA	85	Row crop.

091407-000 INGECO

CODE	MODEL	SCALE	MATER.	MANU.	STOCK #	CTRY.	YEAR	REMARKS
☐ 091407-001	AJ 1.5 HP	1/15	WM KIT	SCALEDOWN MODELS	E2A	GB	??	Small portable gasoline engine.

091420-000 INTERNATIONAL   NOT INTERNATIONAL HARVESTER

CODE	MODEL	SCALE	MATER.	MANU.	STOCK #	CTRY.	YEAR	REMARKS
☐ 091420-001	FORKLIFT		T	ROSKO	04300		??	Industrial forklift.

091909-000 ISEKI

CODE	MODEL	SCALE	MATER.	MANU.	STOCK #	CTRY.	YEAR	REMARKS
☐ 091909-001	TB-20	7"	P	T-N	230	J	??	WFE; Friction motor; With trailer; Formerly listed as a Porsch.

100302-000 JCB

CODE	MODEL	SCALE	MATER.	MANU.	STOCK #	CTRY.	YEAR	REMARKS
☐ 100302-001	3 CX		D	NZG	216	D	81	Industrial backhoe with loader.
☐ 100302-002A	(110B) CRAWLER-LOADER	1/40	D	CORGI	C1110	GB	76	Industrial crawler-loader; Yellow and orange color variations.
☐ 100302-003	870	1/50	D	NZG	141	D	74	Industrial hydraulic excavator.
☐ 100302-004	418	1/35	D	NZG	142	D	75	Industrial four-wheel articulated loader.
☐ 100302-005	5-C	1/32	D & P	BRITAINS LTD.	9580	GB	??	Industrial hydraulic excavator.
☐ 100302-006A	3 CII	1/35	D	NZG	105	D	72	Industrial tractor with backhoe and loader.
☐ 100302-006B	3 CIII	1/35	D	NZG	1058	D	??	Industrial tractor with backhoe and loader.
☐ 100302-007	3 DII	1/35	D	TOMICA	F01	J	81	Industrial tractor with backhoe and loader.
☐ 100302-008	3 D	1/148	WM	STYREX		GB	??	Industrial tractor with backhoe and loader.
☐ 100302-009	3 CII	1/35	D	NZG	109	D	??	Same as 100302-006 but without backhoe; Dark yellow.
☐ 100302-010	520.4 LOADALL	1/32	D & P	BRITAINS LTD.	9519	GB	84	Industrial loader; No. 9814 in the Autoway Series. No. 9519 in the regular series.

100400-000 JOHN DEERE   SEE ALSO DAIN

CODE	MODEL	SCALE	MATER.	MANU.	STOCK #	CTRY.	YEAR	REMARKS
☐ 100400-001	(D)	1/16	CI	VINDEX		USA	30	WFE; Standard style; Plated driver; Rare.
☐ 100400-002	GAS ENGINE	1/16	CI	VINDEX	79	USA	30	Portable gasoline engine.

	CODE	MODEL	SCALE	MATER.	MANU.	STOCK #	CTRY.	YEAR	REMARKS
☐	100400-003	(A)	1/16	CI	ARCADE		USA	41	Row crop; Plated driver.
☐	100400-004	(D)	5"	L	KANSAS TOY	NONE	USA	32	WFE; Standard style; Rare.
☐	100400-005	FROELICH	1/16	CUS.	CHARLES COX	Serial No.	USA	73	WFE; Represents the forerunner of the John Deere tractors; Very limited production.
☐	100400-006A	WATERLOO BOY (N)	1/16	CUS.	CHARLES COX	NONE	USA	74	WFE; Represents a 1914 tractor. Very limited production.
☐	100400-006B	WATERLOO BOY (R)	1/16	CUS.	CHARLES COX	Serial No.	USA	74	WFE; Represents a 1914 tractor. Very limited production.
☐	100400-007	(D)	1/16	CA	OTT; PTW	NONE	USA	69	WFE; Reproduction of Vindex (D) 100400-001.
☐	100400-008	GAS ENGINE	1/16	CA	OTT; PTW	NONE	USA	69	Reproduction of Vindex gasoline engine 100400-002.
☐	100400-009A	(A-GP)	1/12	CA or K	ROBERT GRAY	NONE	USA	71	Row crop; "Steel" wheels.
☐	100400-009B	(A-GP)	1/12	CA	OLD TIME COLLECTABLES	NONE	USA	80	Row crop; "Steel" wheels; Robert Gray molds used to produce this tractor.
☐	100400-009C	(A-GP)	1/12	CA	OLD TIME COLLECTABLES	NONE	USA	84	Row crop; Rubber tires; Decals.
☐	100400-010	(MT)	1/20	P	AUBURN RUBBER		USA	51	Row crop; Molded in driver; "Auburn" name on one side.
☐	100400-011	(A)	1/16	D	LINCOLN SPECIALITIES	NONE	CDN	50	Row crop; Molded in driver; Firestone tires.
☐	100400-012	FROELICH	1/64	D	ERTL	1301	USA	67	WFE; Represents an 1892 model.
☐	100400-013	WATERLOO BOY	1/64	D	ERTL	1302	USA	67	WFE; Represents a 1914 model.
☐	100400-014	(D)	1/64	D	ERTL	1303	USA	67	WFE; Represents a 1932 model.
☐	100400-015	(A)	1/64	D	ERTL	1304	USA	67	Row crop; Represents a 1939 model.
☐	100400-016	(60)	1/64	D	ERTL	1305	USA	67	Row crop; Represents a 1939 model.
☐	100400-017A	(730)	1/64	D	ERTL	1006	USA	67	Row crop; Represents a 1958 model; Metal fenders.
☐	100400-017B	730	1/64	D	ERTL	1306	USA	67	Row crop; Represents a 1958 model; Plastic fenders; "730" decals.
☐	100400-018A	(4010)	1/64	D	ERTL	1307	USA	67	Row crop; Represents a 1960 model; Metal fenders.
☐	100400-018B	(4010)	1/64	D	ERTL	1307	USA	67	Row crop; Represents a 1960 model; Plastic fenders.
☐	100400-019	(4430)	1/64	D	ERTL	1308	USA	72	WFE; Cab; Wheel variations.
☐	100400-020	(D)	1/16	D	ERTL	500	USA	70	WFE; Standard style model; No driver; "Steel" wheels.
☐	100400-021A	(A)	1/16	CA	ERTL		USA	45	Row crop; Molded in driver; Open flywheel; Cast aluminum wheels/tires.
☐	100400-021B	(A)	1/16	CA	ERTL		USA	46	Row crop; Same except rubber wheels/tires; Headlights.
☐	100400-022A	(A)	1/16	CA	ERTL		USA	47	Row crop; Molded in driver; Closed flywheel; Smooth front tires.
☐	100400-022B	(A)	1/16	D	ERTL	NONE	USA	47	Row crop; Same except "ERTL TOY" stamped on side; Arcade front tires.
☐	100400-023A	(B)	1/16	D	ERTL	NONE	USA	50	Row crop; No driver; Metal wheels/rubber tires; Also referred to as "Hi-post A".
☐	100400-023B	(B)	1/16	D	ERTL	NONE	USA	50	Row crop; Same except red.
☐	100400-024	(60)	1/16	D	ERTL		USA	52	Row crop; Light on back of seat; Steerable.
☐	100400-025	110	1/16	D	ERTL	538	USA	65	Lawn and garden tractor; Variations include no 110 decal and half of seat yellow; 110 decal and half of seat yellow; 110 decal with all yellow seat.
☐	100400-026A	140	1/16	D	ERTL	550	USA	67	Lawn and garden tractor; Earlu variation has green grill and metal steering wheel; Late variation has black grill and plastic steering wheel.
☐	100400-026B	140	1/16	D	ERTL	571	USA	69	Lawn and garden tractor; Sunset orange.
☐	100400-026C	140	1/16	D	ERTL	572	USA	69	Lawn and garden tractor; Spruce blue.
☐	100400-026D	140	1/16	D	ERTL	573	USA	69	Lawn and garden tractor; April yellow.
☐	100400-026E	140	1/16	D	ERTL	574	USA	69	Lawn and garden tractor; Patio red.
☐	100400-026F	140	1/16	D	ERTL	515	USA	74	Lawn and garden tractor/blade and trailer; Variations include one lever instead of two and black or green grill.
☐	100400-027A	(400)	1/16	D	ERTL	591	USA	75	Lawn and garden tractor.
☐	100400-027B	(400)	1/16	D	ERTL	515	USA	75	Lawn and garden tractor/blade and trailer.
☐	100400-027C	(400)	1/16	D	ERTL		USA	??	Lawn and garden tractor; "Strobe" decals.
☐	100400-027D	(400)	1/16	D	ERTL	598	USA	??	Lawn and garden tractor; Same except with trailer and blade.
☐	100400-028A	(430)	1/16	D	ERTL	20	USA	62	WFE; Utility; Three-point-hitch.
☐	100400-028B	(430)	1/16	D	ERTL	20	USA	62	WFE; Utility; No three-point-hitch.
☐	100400-029A	(620)	1/16	D	ERTL		USA	56	Row crop; No three-point-hitch; Light on rear of seat.

CODE	MODEL	SCALE	MATER.	MANU.	STOCK #	CTRY.	YEAR	REMARKS
☐	100400-029B (620)	1/16	D	ERTL		USA	56	Row crop; Three-point-hitch; No light on rear of seat.
☐	100400-030A (630)	1/16	D	ERTL	10	USA	58	Row crop; Three-point-hitch; Some had 430 style decal on hood.
☐	100400-030B (630)	1/16	D	ERTL	10	USA	59	Row crop; No three-point-hitch; Either smooth or ribbed front tires.
☐	100400-030C (630)	1/16	D	ERTL	10	USA	60	Row crop; No three-point-hitch; With or without muffler; IH style rear wheels; Also sold in red.
☐	100400-031A (2030)	1/16	D	ERTL	584	USA	73	WFE; Utility.
☐	100400-031B (2030)	1/16	D	ERTL	592	USA	75	WFE; Utility/front loader.
☐	100400-032A (2040)	1/16	D	ERTL	516	USA	76	WFE; Utility; New style front.
☐	100400-032B (2040)	1/16	D	ERTL	517	USA	76	WFE; Utility/loader.
☐	100400-033A 3010-4010	1/16	D	ERTL	530	USA	61	Row crop; Small metal front and rear wheels; Three-point-hitch; Large PTO lever; No fuel filters; One transmission filter on left side of bottom; Generator; Fenders screwed on; NOTE - All 3010-4010 tractors have both front and rear rubber tires; Fenders could be removed to mount corn picker.
☐	100400-033B 3010-4010	1/16	D	ERTL	530	USA	63	Row crop; Same but without three-point-hitch.
☐	100400-034C 3020-4020	1/16	D	ERTL	530	USA	64	Row crop; Small metal front and rear wheels; Three-point-hitch; Three hydraulic and one shift levers; Small levers on both sides of cowl; Small fuel filters on left side; Two transmission filters; Generator; Fenders screwed on; NOTE - All 3020-4020 tractors have rubber rear tires.
☐	100400-034D 3020-4020	1/16	D	ERTL	530	USA	64	Row crop; Same except no three-point-hitch; Bottom three-point holes are open, but no top hole.
☐	100400-034E 3020-4020	1/16	D	ERTL	530	USA	64	Row crop; Small front plastic wheels and metal rear wheels; Three-point-hitch; Three hydraulic and one shift levers; Small lever on each side of cowl; Small fuel filters on left side; Two transmission filters; Generator; Fenders screwed on.
☐	100400-034F 3020-4020	1/16	D	ERTL	530	USA	64	Row crop; Same except without three-point-hitch; Bottom three-point holes open but no top hole.
☐	100400-034G 3020-4020	1/16	D	ERTL	530	USA	64	Row crop; Small plastic front wheels and tires and plastic rear wheels; Three-point-hitch; Three hydraulic and one shift levers; Small lever on each side of cowl; Small fuel filters; Two transmission filters; Generator; Fenders screwed on.
☐	100400-034H 3020-4020	1/16	D	ERTL	530	USA	64	Row crop; Same except without three-point-hitch; Bottom three-point holes open but no top hole; NOTE - No three-point-hitches on later tractors.
☐	100400-034I 3020-4020	1/16	D	ERTL	530	USA	64	Row crop; Small rear tires; Extra large wide rubber front tires; Small metal rear wheels; Three hydraulic and one shift levers; Small lever on each side of cowl; Small fuel filters on left side; Two transmission filters; Generator; Fenders screwed on.
☐	100400-034J 3020-4020	1/16	D	ERTL	530	USA	64	Row crop; Small plastic front wheels and tires and plastic rear wheels; Two hydraulic and one shift levers; Small lever on each side of cowl; Small fuel filters on left side; Two transmission filters; Generator; Fenders screwed on; NOTE - All tractors with small fuel filters will have closed axle housings on the bottom while tractors with large filters will have open axle housings; (As viewed from the bottom).
☐	100400-035A 3020-4020	1/16	D	ERTL	530	USA	64	Row crop; Small front tires and wheels and small rear wheels; One hydraulic and one shift levers; Large fuel filters on left side; No levers on sides of cowl; Two transmission filters; Alternator; Fenders riveted on; No three-point holes.

	CODE	MODEL	SCALE	MATER.	MANU.	STOCK #	CTRY.	YEAR	REMARKS
☐	100400-035B	3020-4020	1/16	D	ERTL	530	USA	64	Row crop; Same except right side three-point hole open.
☐	100400-035C	3020-4020	1/16	D	ERTL	530	USA	64	Row crop; Same except both side three-point holes open.
☐	100400-035D	3020-4020	1/16	D	ERTL	530	USA	64	Row crop; Small front plastic wheels and tires and small rear plastic wheels; No hydraulic or shift levers; Large fuel filters on left side; No side levers on cowl; Two transmission filters; Alternator; No three-point holes.
☐	100400-035E	3020-4020	1/16	D	ERTL	530	USA	64	Row crop; Same except small plastic front tires and wheels and small metal rear wheels.
☐	100400-035F	3020-4020	1/16	D	ERTL	530	USA	65	Row crop; Same except extra large front tires and wide rubber rear tires.
☐	100400-035G	3020-4020	1/16	D	ERTL	530	USA	64	Row crop; Same except extra large plastic front tires and plastic rear wheels.
☐	100400-035H	3020-4020	1/16	D	ERTL	530	USA	65	Row crop; Large plastic front tires and wheels and large rear tires; One hydraulic and one shift levers; Large fuel filters on left side; No side levers on cowl; Two transmission filters; Alternator; Fenders riveted on; No three-point holes.
☐	100400-035I	3020-4020	1/16	D	ERTL	541	USA	65	Row crop; Same except one lever and top three-point hole open.
☐	100400-035J	3020-4020	1/16	D	ERTL	541	USA	65	Row crop; Same except both lower three-point holes open and no top hole.
☐	100400-035K	3020-4020	1/16	D	ERTL	541	USA	65	WFE; Large plastic front wheels and tires and plastic rear wheels; Three hydraulic and one shift levers; Small levers on each side of cowl; Small fuel filters on left side; Two transmission filters; Generator; Fenders screwed on; No three-point holes.
☐	100400-035L	3020-4020	1/16	D	ERTL	553	USA	68	WFE; Large plastic front wheels and tires and plastic rear wheels; Three hydraulic and one shift levers; Small levers on each side of cowl; Small fuel filters on left side; Two transmission filters; Generator; Fenders and ROPS riveted on; Left three-point hole open.
☐	100400-035M	3020-4020	1/16	D	ERTL	547	USA	67	Row crop; Large plastic front wheels and tires and plastic rear wheels; One hydraulic and one shift levers; Large fuel filters on left side; Two transmission filters; Fenders riveted on; Left three-point hole open.
☐	100400-035N	3020-4020	1/16	D	ERTL	530	USA	67	Row crop; Same except both lower three-point holes open.
☐	100400-035O	3030-4020	1/16	D	ERTL	553	USA	68	WFE; Large plastic front wheels and tires and plastic rear wheels; No hydraulic lever; One shift lever; Large fuel filters on left side; Two transmission filters; Large muffler; Alternator; Fenders and ROPS riveted on; Left three-point hole open; SMV sign on rear.
☐	100400-035P	3020-4020	1/16	D	ERTL	553	USA	68	WFE; Same except one shift lever and no three-point holes.
☐	100400-035Q	3020-4020	1/16	D	ERTL	553	USA	68	WFE; Same except one shift lever; Left three-point hole open. NOTE - This list is not necessarily in chronological order.
☐	100400-036A	(4430)	1/16	D	ERTL	512	USA	72	WFE; Sound Gard cab; Fuel and radiator caps; Variations include either small or large lettering on side decals.
☐	100400-036B	(4430)	1/16	D	ERTL		USA	72	Row crop; No cab; Fuel and radiator caps; Very few made.
☐	100400-036C	(4430)	1/16	D	ERTL	521	USA	78	WFE; Cab; No fuel and radiator caps; Small letters on side decals.
☐	100400-036D	(4440)	1/16	D	ERTL		USA	79	WFE; Cab; Strobe decals.
☐	100400-036E	(4440)	1/16	D	ERTL	542	USA	79	WFE; Cab; Dual rear wheels; Front spindle variations, large or small.
☐	100400-036F	(4440)	1/16	D	ERTL	542	USA	80	WFE; Cab; Dual rear wheels; Strobe decals; Large front rubber tires.
☐	100400-036G	4250 TOY FARMER	1/16	D	ERTL	5507AA	USA	82	WFE; Front-wheel-assist; Cab; Dual rear wheels; Air cleaner; Only 1550 made; "NATIONAL FARM TOY SHOW" tractor.

	CODE	MODEL	SCALE	MATER.	MANU.	STOCK #	CTRY.	YEAR	REMARKS
☐	100400-036H	(4450)	1/16	D	ERTL	5506	USA	83	WFE; Cab; Strobe decals including light bar on front; Air cleaner.
☐	100400-036I	(4450)	1/16	D	ERTL	5507DO	USA	83	WFE; Cab; Strobe decals including light bar on front; Air cleaner; Dual rear wheels.
☐	100400-037A	(4430)	1/32	D	ERTL	66	USA	73	Row crop; Cab.
☐	100400-037B	(4440)	1/32	D	ERTL	66	USA	79	Row crop; Cab; Strobe decals.
☐	100400-037C	(4440)	1/32	D	ERTL	66	USA	80	Row crop; Strobe decals; No cab; Very few made.
☐	100400-037D	50 SERIES	1/32	D	ERTL	66	USA	83	Row crop; Same except having light bar on top, front of hood.
☐	100400-038	4430	1/25	PK	ERTL	8005	USA	73	WFE; Cab.
☐	100400-039A	5020	1/16	D	ERTL	555	USA	69	WFE; "Wheatland" style standard tractor; Air cleaner; No front axle braces.
☐	100400-039B	5020	1/16	D	ERTL	555	USA	??	WFE; Same except no air cleaner.
☐	100400-039C	5020	1/16	D	ERTL	555	USA	??	WFE; Same except with front axle braces; Two piece air cleaner.
☐	100400-039D	5020	1/16	D	ERTL	555	USA	??	WFE; Same except solid front axle braces.
☐	100400-039E	5020	1/16	D	ERTL	555	USA	??	WFE; Same except with long side decals; No air cleaner.
☐	100400-040A	(7520)	1/16	D	ERTL	510	USA	72	Four-wheel-drive; Cab; Articulated; Air cleaner.
☐	100400-040B	(7520)	1/16	D	ERTL	510	USA	75	Four-wheel-drive; Cab; Articulated; No air cleaner.
☐	100400-041A	(8630)	1/16	D	ERTL	597	USA	75	Four-wheel-drive; Cab; Articulated; Dual wheels.
☐	100400-041B	(8640)	1/16	D	ERTL	597	USA	79	Four-wheel-drive; cab; Articulated; Dual wheels; Strobe decals; Some of the later models have small lower cab windows, apparently the wrong cab having been installed.
☐	100400-041C	(8650)	1/16	D	ERTL	5508CO	USA	83	Four-wheel-drive; Cab; Articulated; Dual wheels; Strobe decals with light bar on front.
☐	100400-042A	(40) CRAWLER	1/16	D	ERTL		USA	54	Crawler; Green; Blade optional.
☐	100400-042B	(40) INDUSTRIAL CRAWLER	1/16	D	ERTL		USA	55	Industrial crawler; Yellow; Blade optional.
☐	100400-042C	(420) CRAWLER	1/16	D	ERTL		USA	56	Crawler; Green with yellow side stripe; Blade optional.
☐	100400-043	(440) INDUSTRIAL CRAWLER	1/16	D	ERTL		USA	59	Industrial crawler; Yellow; Black embossed grill; No blade.
☐	100400-044A	(440) INDUSTRIAL TRACTOR	1/16	D	ERTL		USA	59	WFE; Industrial wheel tractor; Yellow/black seat. Three-point-hitch.
☐	100400-044B	(1010) INDUSTRIAL TRACTOR	1/16	D	ERTL		USA	62	WFE; Industrial utility tractor; Yellow; No three-point-hitch.
☐	100400-045	(1010) INDUSTRIAL CRAWLER	1/16	D	ERTL	526	USA	63	Industrial crawler; Yellow; With blade.
☐	100400-046A	(JD-450) INDUSTRIAL CRAWLER	1/16	D	ERTL	546	USA	65	Industrial crawler/blade. Two levers on right side of seat; Black radiator decal.
☐	100400-046B	(JD-450) INDUSTRIAL CRAWLER	1/16	D	ERTL	554	USA	67	Industrial crawler/blade and winch; No levers on right side; Silver radiator decals.
☐	100400-046C	(JD-450) INDUSTRIAL CRAWLER	1/16	D	ERTL	509	USA	73	Industrial crawler/blade, winch and ROPS.
☐	100400-047	(JD-350)	1/16	D	ERTL	589	USA	75	WFE; Industrial tractor/backhoe and loader; Front wheel variations include yellow or black inside.
☐	100400-048	(JD-310)	1/25	PK	ERTL	8015	USA	75	WFE; Industrial tractor/backhoe and loader.
☐	100400-049A	644 FOUR-WHEEL-LOADER	1/25	D	ERTL	503	USA	71	Industrial four-wheel-loader; Articulated.
☐	100400-049B	644 FOUR-WHEEL-LOADER	1/25	D	ERTL	507	USA	73	Industrial four-wheel-loader/ROPS; Articulated.
☐	100400-050	(570) GRADER	1/25	D	ERTL	1/25	USA	71	Industrial motor grader/cab; Articulated.
☐	100400-051	(690) HYDRAULIC EXCAVATOR	1/25	D	ERTL	505	USA	71	Industrial hydraulic excavator/crawler tracks; Cab.
☐	100400-052A	(860) SCRAPER	1/25	D	ERTL	506	USA	71	Industrial elevating scraper pan.
☐	100400-052B	(869) SCRAPER	1/25	D	ERTL	507	USA	73	Industrial elevating scraper pan/ROPS.
☐	100400-053	LOG SKIDDER	1/16	D	ERTL	590	USA	75	Industrial four-wheel-drive; Articulated; With blade, winch and ROPS.
☐	100400-054	2020	1/16	P	WADER		D	67	WFE; Utility; Rare.
☐	100400-055	2120	1/43	P	NACORAL	2120	E	73	WFE; Utility; Green with silver plating; Rare.
☐	100400-056A	(3020)	1/32	D	LEE TOYS	9716-1881	USA	??	Row crop; Red and silver.
☐	100400-056B	(3020)	1/32	D	LEE TOYS	1135-1891	USA	??	Row crop; Red, silver and brown; Cab; Dual rear wheels.

	CODE	MODEL	SCALE	MATER.	MANU.	STOCK #	CTRY.	YEAR	REMARKS
☐	100400-057A	JD-LANZ 700	1/61	D	LESNEY	508	GB	63	WFE; Utility; Gray tires.
☐	100400-057B	JD-LANZ 700	1/61	D	LESNEY	50	GB	65	WFE; Utility; Black tires.
☐	100400-058	JD-LANZ	1/61	D	TINY CAR		BR	68	WFE; Utility.
☐	100400-059A	JD-LANZ (300/500)	5"	P	REX		D	62	WFE; Utility; One variation has headlights, one does not; Rare.
☐	100400-059B	JD-LANZ (300/500)	5"	P	REX		D	64	WFE; Utility; No headlights; Rare.
☐	100400-060	JD-LANZ 1010	5-1/2"	P	REX		D	62	Industrial crawler/loader; Rare.
☐	100400-061A	SKID STEER LOADER	1/16	D	ERTL	569	USA	77	Four-wheel skid steer loader; Green; Some had white wheels.
☐	100400-061B	SKID STEER LOADER	1/16	D	ERTL	6571	USA	81	Industrial four-wheel skid steer loader; Yellow.
☐	100400-062A	(4230)	1/64	D	SIGOMEC		RA	78	WFE; No cab; Green; Similar to Ertl John Deere (100400-019).
☐	100400-062B	(4230)	1/64	D	SIGOMEC		RA	78	WFE; Industrial tractor; No cab; Yellow with black radiator screen and seat.
☐	100400-063A	(4230)	1/16	D	SIGOMEC		RA	78	WFE; No cab; Three-point-hitch; Similar to Ertl John Deere (100400-036).
☐	100400-063B	(4230)	1/16	D	SIGOMEC		RA	78	WFE; Industrial tractor; Yellow; No cab; No three-point-hitch; Black radiator screen and seat.
☐	100400-064	A - GP	1/16	CUS.	DENNIS PARKER	NONE	USA	79	Row crop; "Steel" wheels; excellent detail; Only 250 made.
☐	100400-065	(4430)	1/64	D	ERTL	1619	USA	78	WFE; Cab; Oscillating front axle.
☐	100400-066A	GAS ENGINE	1/16	K	ROBERT GRAY	NONE	USA	81	Portable gasoline engine.
☐	100400-066B	GAS ENGINE	1/16	K	ROBERT GRAY	NONE	USA	81	Same but with wheels.
☐	100400-067	WATERLOO BOY	1/16	CA	SCALE MODELS	NO. 3	USA	82	WFE; Represents a 1916 model; No. 3 in the Threshers Series; "Steel" wheels.
☐	100400-068A	(A)	1/87	WM KIT	WOODLANDS SCENICS	D-211	USA	81	Row crop on "steel".
☐	100400-068B	(A)	1/87	WM KIT	WOODLANDS SCENICS	D-208	USA	81	Row crop on rubber.
☐	100400-069	(A)	1/87	WM KIT	WOODLANDS SCENICS	M112	USA	81	Row crop; Part of a set.
☐	100400-070	(A)	1/16	D	SCALE MODELS	NO. 7	USA	81	Row crop on "steel"; No. 7 in the JLE Collector Series; 3000 made.
☐	100400-071A	B	1/32	WM KIT	BROWN'S MODELS		GB	81	WFE: Unstyled.
☐	100400-071B	B	1/32	WM KIT	BROWN'S MODELS		GB	81	Row crop/single front wheel: Unstyled.
☐	100400-071C	B	1/32	WM KIT	BROWN'S MODELS		USA	81	Row crop/dual front wheels: Unstyled.
☐	100400-071D	B	1/32	WM KIT	BROWN'S MODELS		GB	82	WFE; Styled.
☐	100400-071E	B	1/32	WM KIT	BROWN'S MODELS		GB	73	Row crop/single front wheel; Styled.
☐	100400-071F	B	1/32	WM KIT	BROWN'S MODELS		GB	83	Row crop/dual front wheels; Styled.
☐	100400-072A	3140	1/32	D	ERTL	H1635	USA	79	WFE; Front-wheel-assist; European cab; Small wheels.
☐	100400-072B	3140	1/32	D	ERTL	H1635	USA	80	WFE; Front-wheel-assist; European cab; Slightly larger wheels.
☐	100400-073	3140	1/32	D	ERTL	5512	USA	82	WFE; Front-wheel-assist; American style Sound Gard cab (rounded).
☐	100400-074	4430	1/64	PK	??		??	??	WFE; Cab.
☐	100400-075	(4440)	1/16	P	ERTL	0031	USA	81	WFE; Cab; Radio controlled.
☐	100400-076	(4440)	1/64	D	ERTL	1619	USA	81	WFE; Cab; Strobe decals.
☐	100400-077A	SNOWMOBILE	1/12	P	NORMATT	7000	USA	72	Snowmobile; Green; No motor push type.
☐	100400-077B	SNOWMOBILE	1/12	P	NORMATT	7000	USA	72	Snowmobile; Green; Battery operated.
☐	100400-078	JDX SNOWMOBILE	1/12	P	NORMATT	7000-X	USA	75	Snowmobile; Black; Battery operated.
☐	100400-079A	440 CYCLONE SNOWMOBILE	1/12	P	SUTTLE	8111	USA	81	Snowmobile; Metallic green and black; No motor push type.
☐	100400-079B	440 CYCLONE SNOWMOBILE	1/12	P	SUTTLE	8111	USA	81	Snowmobile; Metallic green and black; Battery operated.
☐	100400-080	TRAILFIRE 440	1/12	P	ERTL	573	USA	81	Snowmobile; Silver-gray; Battery operated.
☐	100400-081	(AR)	1/16	CA	A T & T COLLECTABLES	NONE	USA	82	WFE; Standard style tractor.
☐	100400-082A	(80)	1/16	CA	ELDON TRUMM	NONE	USA	82	WFE; Standard style tractor; Wheel variations.
☐	100400-082B	(80) INDUSTRIAL	1/16	CA	ELDON TRUMM	NONE	USA	82	WFE; Same except industrial yellow.
☐	100400-082C	(820)	1/16	CA	ELDON TRUMM	NONE	USA	82	WFE; Standard style tractor; Green with yellow side stripe; Wheel variations.
☐	100400-082D	(820) INDUSTRIAL	1/16	CA	ELDON TRUMM	NONE	USA	82	WFE; Same except industrial yellow.
☐	100400-083A	(830) DIESEL	1/16	CA	ELDON TRUMM	NONE	USA	83	WFE; Standard style tractor; Green with wide yellow side stripe.
☐	100400-083B	(830) DIESEL INDUSTRIAL	1/16	CA	ELDON TRUMM	NONE	USA	83	WFE; Same except industrial yellow.

	CODE	MODEL	SCALE	MATER.	MANU.	STOCK #	CTRY.	YEAR	REMARKS
☐	100400-084	(R) DIESEL	1/16	CA	ELDON TRUMM	NONE	USA	83	WFE; Standard style tractor.
☐	100400-085A	(950) COMPACT UTILITY	1/16	D	ERTL	581	USA	82	WFE; Small compact utility; ROPS; Front-wheel-assist; Recall version had no "wedge" to hold muffler in place.
☐	100400-085B	(950) COMPACT UTILITY	1/16	D	ERTL	581	USA	82	WFE; Same except with "wedge".
☐	100400-086	(GP)	1/16	CA	SCALE MODELS	NO. 9	USA	82	Row crop; Sold as "Green" tractor; No. 9 in the JLE Collector Series; 3000 made.
☐	100400-087A	8650 COL. SER.	1/16	D	ERTL	5508	USA	82	Four-wheel-drive; Cab; Articulated; Dual Wheels; Muffler and air cleaner set off to one side; "COLLECTOR SERIES JULY 1982".
☐	100400-087B	(8650)	1/16	D	ERTL	5508	USA	82	Four-wheel-drive; Same but without collector inscription.
☐	100400-088A	4850 COL. SER.	1/16	D	ERTL	584DA	USA	82	WFE; Front-wheel-assist; Cab; Dual rear wheels; "COLLECTOR SERIES NEW ORLEANS 7/82".
☐	100400-88B	(4850)	1/16	D	ERTL	584	USA	82	WFE; Same except without collector inscription.
☐	100400-089A	(4440)	1 Liter	POR	PACESETTER		USA	83	WFE; "GREEN MACHINE"; 4200 made with John Deere logo; Decanter.
☐	100400-089B	(4440)	1 Liter	POR	PACESETTER		USA	83	WFE; "GREEN MACHINE" 1800 made with Pacesetter logo; Decanter.
☐	100400-090	(4440)	200 ML	POR	PACESETTER		USA	83	WFE; "LITTLE GREEN MACHINE" 3600 made.
☐	100400-091A	(M)	1/16	CUS.	K & G SAND CASTING		USA	83	WFE; Represents a 1947-52 model.
☐	100400-091B	(MT)	1/16	CUS.	K & G SAND CASTING	NONE	USA	83	Row crop; represents a 1949-52 model.
☐	100400-092	(MC)	1/16	CUS.	K & G SAND CASTING	NONE	USA	83	Crawler; Represents a 1949-52 model.
☐	100400-093	(G)	1/16	CUS.	PETE FREIHEIT	NONE	USA	83	Row crop; "Roll-o-matic" front axle; Excellent detail.
☐	100400-094A	(G) COL. SER.	1/16	D	SCALE MODELS	NO. 1	USA	83	Row crop on "steel" No. 1 in the JLE Collector Series II; Represents a 1942 model; 5000 made.
☐	100400-094B	(G)	1/16	D	SCALE MODELS		USA	83	Row crop on rubber; "1983 DYERSVILLE SHOW" inscription.
☐	100400-095A	2550 COL. SER.	1/16	D	ERTL	501DA	USA	83	WFE; Front-wheel-assist; Cab; Utility; COLLECTOR SERIES.
☐	100400-095B	(2550)	1/16	D	ERTL	501DO	USA	83	WFE; Front-wheel-assist; Utility; Cab.
☐	100400-095C	(2550)/LDR.	1/16	D	ERTL	503DO	USA	83	WFE; Front-wheel-assist; Cab; Utility; With front loader.
☐	100400-096	(8450)	1/64	D	ERTL	5509FO	USA	83	WFE; Cab; Black exhaust and air cleaner.
☐	100400-097	(8550)	1/64	D	ERTL	575EO	USA	84	WFE; Four-wheel-drive; Cab; Dual wheels; Articulated.
☐	100400-098	BACKHOE	1/32	D	ERTL	5520DO	USA	83	Industrial backhoe-loader.
☐	100400-099	BACKHOE	1/64	D	ERTL	5521EO	USA	83	Industrial backhoe-loader.
☐	100400-100A	GP ROW CROP	1/16	CA	SCALE MODELS	NONE	USA	84	Row crop on "steel"; "3-10-84" on frame; Introduced at the 1984 Lafayette Show.
☐	100400-100B	GP ROW CROP	1/16	CA	SCALE MODELS		USA	84	Row crop on "steel"; Rockford, Illinois show tractor; Has exhaust and air cleaner stacks; No fenders; "6-17-84" on frame.
☐	100400-101	(H)	1/16	CUS.	LYLE DINGMAN	NONE	USA	84	Row crop; Variations include open flywheel or electric starter and with or without headlights and fenders.
☐	100400-102A	7020	1/32	CA KIT	ROY LEE BAKER	NONE	USA	84	Four-wheel-drive; Articulated; With or without cab.
☐	100400-102B	7020	1/32	CA	ROY LEE BAKER	NONE	USA	84	Four-wheel-drive; Articulated; With or without cab.
☐	100400-103	8010	1/16	CUS.	DAVID SHARP	NONE	USA	83	Four-wheel-drive; Articulated.
☐	100400-104A	3185	1/43	D & P	NPS	3009	HK	84	WFE; Front-wheel-assist; Green; Gyro motor.
☐	100400-104B	3185	1/43	D & P	NPS	3009	HK	84	WFE; Front-wheel-assist; Industrial yellow and black; Gyro motor.
☐	100400-105	(D)	1/16	CUS.	DENNIS PARKER	NONE	USA	83	WFE; Rubber tires; Ertl "D" customized (100400-020).
☐	100400-106A	(A) COL. SER.	1/16	D	ERTL	538DO	USA	84	Row crop on "steel" Commemorates the 50th anniversary of the John Deere "A".
☐	100400-106B	(A)	1/16	D	ERTL	538	USA	84	Row crop on rubber; Represents a 1934 model; Exhaust and air cleaner slightly different; No fenders.
☐	100400-107A	50 SERIES	1/16	D	ERTL	541DO	USA	84	WFE; Black cab posts and green tp; Different engine detail; Plastic cab interior incuding high back style seat, floor and dash panel; Top three point hitch link cast in; Light bar on front decal.

CODE	MODEL	SCALE	MATER.	MANU.	STOCK #	CTRY.	YEAR	REMARKS
100400-108	50 SERIES ROW CROP	1/64	D	ERTL	5509FO	USA	84	WFE; Cab; Front weights; Black cab posts.
100400-109	4020	1/16	CA KIT	DENNIS PARKER	NONE	USA	84	WFE; Available only as body castings kit to be assembled.
100400-110	WATERLOO BOY	O Scale	WM KIT	DON WINTER	NONE	USA	81	WFE; Represents an antique style tractor.
100400-111A	8010	1/16	CA	ELDON TRUMM	NONE	USA	85	WFE; Four-wheel-drive; Articulated; Represents a 1959 model.
100400-111B	8020	1/16	CA	ELDON TRUMM	NONE	USA	85	WFE; Four-wheel-drive; Articulated; Updated version of the 8010; Represents a 1961 model.
100400-112	D	1/10	WOOD	KRUSE	NONE	USA	78	WFE; "Steel" wheels; Spoke flywheel.
100400-113	GP STANDARD	1/10	WOOD TREAD	KRUSE	NONE	USA	78	WFE; "Steel" wheels.
100400-114	GP WIDE TREAD	1/10	WOOD	KRUSE	NONE	USA	78	Row crop; "Steel" wheels.
100400-115	A	1/10	WOOD	KRUSE	NONE	USA	78	Row crop; "Steel" wheels.
100400-116	B	1/10	WOOD	KRUSE	NONE	USA	80	Row crop; "Steel" wheels.
100400-117	G	1/10	WOOD	KRUSE	NONE	USA	79	Row crop; "Steel" wheels.
100400-118	A W STYLED	1/10	WOOD	KRUSE	NONE	USA	82	Row crop; Rubber.
100400-119	A STYLED	1/10	WOOD	KRUSE	NONE	USA	82	Row crop; Rubber.
100400-120	B STYLED	1/10	WOOD	KRUSE	NONE	USA	82	Row crop; Rubber.
100400-121	G STYLED	1/10	WOOD	KRUSE	NONE	USA	82	Row crop; Rubber.
100400-122	D STYLED	1/10	WOOD	KRUSE	NONE	USA	81	WFE; Rubber.
100400-123A	L STYLED	1/10	WOOD	KRUSE	NONE	USA	82	WFE; Rubber.
100400-123B	L INDUSTRIAL	1/10	WOOD	KRUSE	NONE	USA	83	WFE; Rubber; Yellow.
100400-124	L UNSTYLED	1/10	WOOD	KRUSE	NONE	USA	83	WFE; Rubber.
100400-125	JOHN DEERE-LINDEMANN	1/10	WOOD	KRUSE	NONE	USA	83	Crawler.
100400-126	50	1/10	WOOD	KRUSE	NONE	USA	83	Row crop; Rubber.
100400-127	60	1/10	WOOD	KRUSE	NONE	USA	84	Row crop; Rubber.
100400-128	80	1/10	WOOD	KRUSE	NONE	USA	82	WFE; Rubber.
100400-129	R	1/10	WOOD	KRUSE	NONE	USA	82	WFE; Rubber.
100400-130	520	1/10	WOOD	KRUSE	NONE	USA	83	Row crop; Rubber.
100400-131	620	1/10	WOOD	KRUSE	NONE	USA	84	Row crop; Rubber.
100400-132	720 GAS	1/10	WOOD	KRUSE	NONE	USA	82	Row crop; Rubber.
100400-133	820	1/10	WOOD	KRUSE	NONE	USA	80	WFE; Rubber.
100400-134	430	1/10	WOOD	KRUSE	NONE	USA	84	WFE; Rubber.
100400-135	530	1/10	WOOD	KRUSE	NONE	USA	83	Row crop; Rubber.
100400-136	630	1/10	WOOD	KRUSE	NONE	USA	84	Row crop; Rubber.
100400-137	730 GAS	1/10	WOOD	KRUSE	NONE	USA	81	Row crop; Rubber.
100400-138	830	1/10	WOOD	KRUSE	NONE	USA	79	WFE; Rubber.
100400-139	D	1/10	WOOD	KRUSE	NONE	USA	81	WFE; "Steel" wheels; In glass case.
100400-141	D UNSTYLED	1/10	WOOD	KRUSE	NONE	USA	82	WFE; Rubber.
100400-142	G	1/10	WOOD	KRUSE	NONE	USA	82	Row crop; Rubber; Cast wheels; Adjustable axle.
100400-143	G	1/10	WOOD	KRUSE	NONE	USA	80	Row crop; Rubber; Spoke wheels.
100400-144	G	1/10	WOOD	KRUSE	NONE	USA	81	Row crop; Rubber; Disc wheels.
100400-145	730	1/10	WOOD	KRUSE	NONE	USA	82	Row crop; With fenders.
100400-146	730 WHEATLAND	1/10	WOOD	KRUSE	NONE	USA	83	WFE; Rubber; Fenders over rear wheels.
100400-147	730	1/10	WOOD	KRUSE	NONE	USA	83	WFE; Rubber.
100400-148	GAS ENGINE	1/10	WOOD	KRUSE	NONE	USA	79	Gasoline engine. Represents the 6 horsepower model.
100400-149	GP	1/16	CA	A T & T COLLECTABLES	NONE	USA	82	Row crop; "Steel" wheels.
100400-150	GP 2-ROW	1/16	CA	A T & T COLLECTABLES	NONE	USA	83	Row crop; "Steel" wheels.
100400-151A	B	1/16	CA	A T & T COLLECTABLES	NONE	USA	83	Row crop; "Steel" wheels.
100400-151B	B	1/16	CA	A T & T COLLECTABLES	NONE	USA	83	Row crop; Rubber; Fenders.
100400-152A	G	1/16	CA	A T & T COLLECTABLES	NONE	USA	83	Row crop; Rubber; Cast wheels.
100400-152B	G	1/16	CA	A T & T COLLECTABLES	NONE	USA	83	Row crop; Rubber; Spoke wheels.
100400-153	A	1/32	P	??	NONE	F	??	Row crop.
100400-154	A	1/16	CUS.	PETE FREIHEIT	NONE	USA	85	Row crop; Steerable; Cast rear wheels.
100400-155	(D)	5-1/2"	D	BANTHRICO		USA	82	WFE; Bank.
100400-156	(A) GP	1/16	CUS.	LUCHT-FRIESEN	NONE	USA	78	Row crop; Separately cast, plated driver, flywheel, pulley and steering post; "Steel" wheels; Very limited production.
100400-157	JD-850	1/43	PEWTER	PRECISION CRAFT PEWTER		USA	84	Industrial crawler with blade and ROPS.
100400-158	430	1/16	CUS.	GEORGE NYGREN	NONE	USA	83	Row crop; Three-point-hitch; Excellent detail; Sold as "Green 430 tractor"; Decals added.
100400-159	430 CRAWLER	1/16	CUS.	GEORGE NYGREN	NONE	USA	83	Crawler; Three-point-hitch; Excellent detail; Sold as "Green 430 crawler"; Decals added.
100400-160	435	1/16	D	SIGOMEC		RA	??	WFE; Utility; Similar to Ertl 430 (100400-028).

319

| --- | --- | --- | --- | --- | --- | --- | --- | --- |
☐	100400-161	730	1/16	D	SIGOMEC		RA	??	Row crop; Similar to Ertl 730 (100400-029).
☐	100400-162	(3010)	1/16	D	SIGOMEC		RA	??	Row crop; Similar to Ertl 3010-3020 series.
☐	100400-163	??	1/16	CA	SIGOMEC		RA	??	WFE; Utility; Also available with loader.
☐	100400-164	PACESETTER (8650)	200 ML	POR	PACESETTER	NO. 1	USA	83	WFE; Four-wheel-drive; Decanter.
☐	100400-165	PACESETTER (8650)	1 Liter	POR	PACESETTER	NO. 1	USA	83	WFE; Four-wheel-drive; Decanter.
☐	100400-166	(A)	1/16	CA	??			????	Row crop; Separate driver; Aluminum wheels/tires; Similar to Ertl "A" (100400-023); Rare.
☐	100400-167	(D)	1/16	CA	EARL JERGESEN	NONE	USA	83	WFE; "Steel"; Air cleaner and exhaust stacks.
☐	100400-168	(A)	1/64	D	??			83	Row crop; Copy of Ertl A (100400-015).
☐	100400-169	PACESETTER (8650)	1 Liter	POR	PACESETTER	NO. 1	USA	83	WFE; Four-wheel-drive; Decanter.

110918-000 KIROVEC

☐	110918-001	??	1/43	D & P	LENINGRAD		USSR	84	WFE; Four-wheel-drive; Cab; Three-point-hitch; Orange or yellow over white.

111801-000 KRAMER

☐	111801-001	??	12 cm	P	CURSOR		D	??	WFE; Seat back rails on both fenders.

112102-000 KUBOTA

☐	112102-001A	(L2200)	1/42	D	TOMICA	92	J	75	WFE.
☐	112102-001B	(L2200)	1/42	D	TOMICA AVIVA	C10	J	78	WFE; Same except with comic character "CHARLIE BROWN" on driver's seat.
☐	112102-002A	L-1500	1/23	D	DIAPET	0386 *	J	75	WFE * 0381 with trailer; 0382 with disk, 0383 with roto-tiller; 0386 tractor only.
☐	112102-002B	L-245	1/23	D	DIAPET	0386 *	J	78	WFE; Same combinations and stock numbers as 112102-002A.
☐	112102-002C	L-1500	1/23	D	NPS	1203	HK	80	WFE; Identical to Play Art Kubota 112102-005.
☐	112102-003	??	1/17	P	KUBOTA-JIGYOSHA		J	79	WFE; "KUBOTA'S PIGGY BANK" Very well detailed tractor mounted on a base which serves as the bank; Tractor is detachable.
☐	112102-004	L2402DT SUNSHINE	1/23	D	DIAPET	127- T-90	J	79	WFE; Front-wheel-assist.
☐	112102-005	KUBOTA (L2402DT)	1/23	D	PLAY ART DICKIE	7595	HK	81	WFE; Copy of Diapet Kubota (112102-002).
☐	112102-006	M79500T	1/20	D	DIAPET	127- T-108	J	84	WFE; Front-wheel-assist; Excellent detail with opening doors and hood panels.

120113-000 LAMBORGHINI

☐	120113-001A	R 1056	1/43	D & P	FORMA-PLAST	.086	I	77	WFE; Front-wheel-assist; Cab; Similar in design as the SAME Buffalo 130 (190113-004A); Marked "Forma Toys".
☐	120113-001B	R 1056	1/43	D & P	YAXON	.086	I	??	WFE; Front-wheel-assist; Same except different manufacturer. Marked "Yaxon".
☐	120113-002A	??	1/16	P	CAVALLINO		I	83	WFE; ROPS; Red, white and blue.
☐	120113-002B	??	1/16	P	CAVALLINO		I	83	WFE; Same except with front loader.
☐	120113-003	(653)	1/50	P	GRISONI	43	I	??	Crawler; Also available with a blade

120114-000 LANDINI   NOW A PART OF MASSEY FERGUSON LTD.

☐	120114-001A	R-4000	1/43	D	MERCURY	523	I	61	WFE; Hood hinged in front; Blue and brown with yellow wheels.
☐	120114-001B	R-5000	1/43	D	MERCURY	523	I	??	WFE; Same except different hood and fenders; Very rare.
☐	120114-002A	12500 or 14500	1/43	D & P	YAXON	063	I	79	WFE; Two-wheel-drive or front-wheel-assist; ROPS or cab; Blue and gray; Grill style #1.
☐	120114-002B	12500 or 14500	1/43	D & P	YAXON	063	I	81	WFE; Two-wheel-drive or front-wheel-assist (DT); ROPS or cab; Blue and gray; Grill style #2.
☐	120114-002C	12500 or 14500	1/43	D & P	YAXON	063	I	83	WFE; Two-wheel-drive or front-wheel-assist (DT); ROPS or cab; Blue and gray; Grill style #3.
☐	120114-003	R5000N SPECIAL	1/15	P	SIACA		I	71	WFE; Very well detailed; Similar to Revell model (120114-004).
☐	120114-004	R5000N SPECIAL	1/15	PK	REVELL		D	71	WFE; Very well detailed.
☐	120114-005	L35 HOT BULB	1/16	P	GIANCO		I	55	WFE; Gray; Clockwork; Rare; One cylinder horizontal hot bulb engine.
☐	120114-006	??	1/43	D	YAXON	501	I	84	Industrial four-wheel-drive; 501 with loader; 502 with loader and backhoe; 503 with snow plow; 504 with blade and backhoe; Identical to Yaxon MF industrial models.

120126-000 LANZ

☐	120126-001	BULLDOG	1/86	P	AUGUPLAS-MINICARS	25	E	59	WFE.
☐	120126-002	BULLDOG	1/43	P	MARKLIN	8002	D	50	WFE; Has "LANZ" name on front; Has exhaust stack; Separate driver.

	CODE	MODEL	SCALE	MATER.	MANU.	STOCK #	CTRY.	YEAR	REMARKS
☐	120126-003	BULLDOG	1/43	D	MARKLIN	8029	D	59	WFE; No exhaust stack; Separate driver; Color variations.
☐	120126-004	BULLDOG	1/21	P	REX		D	59	WFE; Blue with red wheels, red with yellow wheels or green with yellow wheels.
☐	120126-005	BULLDOG	1/87	P	WIKING	V-308	D	73	WFE; Represents old style tractor.
☐	120126-006	BULLDOG	1/43	D	MARKLIN	8022/81	D	39	WFE; Has "LANZ" name on front; With exhaust stack; Clockwork; Color variations.
☐	120126-007	BULLDOG			JAYA	853	E	65	WFE; Orange; Made under license by Lanz Iberica S.A.; Name "LANZ IBERICA" on the flywheel.
☐	120126-008	BULLDOG	1/200	D	MERCATOR		D	73	WFE; Military green.
☐	120126-009	55 PS	1/43	WM KIT	STEAM & TRUCK		GB	??	WFE; Tractor-car; Only 500 made.
☐	120126-010	24PS	1/32		??			??	
☐	120126-011	BULLDOG	1/87		??			??	
☐	120126-012	EILBULLDOG	1/43	WM	STEAM & TRUCK	ST01	D	82	WFE; Road tractor; EIL means FAST.
☐	120126-013A	BULLDOG 3506	1/43	WM	STEAM & TRUCK	ST02	D	84	WFE.
☐	120126-013B	BULLDOG	1/43	WM	STEAM & TRUCK	ST03	D	84	WFE; Industrial version.
☐	120126-014	12 HP 1921	1/43	WM	STEAM & TRUCK	ST06	D	84	WFE.
☐	120126-015	BULLDOG	5"	P	APTHKYA		USSR	73	WFE; Formerly listed as Kohctpykto.
☐	120126-016	12 PS 1921	1/43	WM KIT	STEAM 7 TRUCK	ST 07	D	84	WFE.

120525-000 LEYLAND

	CODE	MODEL	SCALE	MATER.	MANU.	STOCK #	CTRY.	YEAR	REMARKS
☐	120525-001A	384	1/43	D	DINKY	308	GB	71	WFE; Metallic blue with white wheels; With driver.
☐	120525-001B	384	1/43	D	DINKY	308	GB	71	WFE; Metallic red with white wheels; With driver.
☐	120525-001C	384	1/43	D	DINKY	308	GB	77	WFE; Straw yellow with white wheels; With driver.
☐	120525-002	(384)	1/12	P	H P PLAST (NYRHINEN)		SF	80	WFE; Blue or white front wheels.
☐	120525-003	804	1/18	C/B	KARRAN PRODUCTS LTD.	J58281	GB	81	WFE; Cardboard kit.
☐	120525-004	802	1/43	D	YAXON	035	I		WFE; Front-wheel-assist; Cab; Straw yellow and black; Yaxon used the Ford casting to produce this model.

120919-000 LISTER

	CODE	MODEL	SCALE	MATER.	MANU.	STOCK #	CTRY.	YEAR	REMARKS
☐	120919-001	TYPE D	1/15	WM KIT	SCALEDOWN MODELS		GB	80	Small portable gasoline engine.

130106-000 MASSEY-FERGUSON

	CODE	MODEL	SCALE	MATER.	MANU.	STOCK #	CTRY.	YEAR	REMARKS
☐	130106-001	(44)	1/43	D	DINKY	300	GB	??	WFE; Similar to Dinky Massey-Harris (44) 130108-001.
☐	130106-002A	??	1/43	D	MICRO-MODELS	8	AUS	58	WFE.
☐	130106-002B	??	1/43	D	MICRO-MODELS	12	AUS	58	Road roller.
☐	130106-003	ERROR							THIS IS ACTUALLY A YANMAR TRACTOR.
☐	130106-004	35	1/43	D	VILMER	575	DK	59	WFE.
☐	130106-005	FERGUSON TE-30	1/43	D	CRESCENT	1203	GB	60	WFE; With clockwork.
☐	130106-006	??	1/40	D	PEETZY-ROCO	U-301	A	??	WFE.
☐	130106-007	35	1/32	D	LION-MOLBERG		DK	??	WFE; Utility; Hood hinged in front and lifts.
☐	130106-008A	35	1/25	D	MORGAN MILTON LTD.	1035	IN	70	WFE; Utility; Excellent detail.
☐	130106-008B	35	1/25	D	MILTAN	1035	IN	83	WFE; Utility; Excellent detail; Slightly different air cleaner and exhaust stack.
☐	130106-009	35	1/43	D	MERCURY	510	I	61	WFE; Utility; With a disk plow, baler and forage wagon.
☐	130106-010	35	1/43	D	GAMDA	30	IL	65	WFE; Utility.
☐	130106-011	(35)	1/32	D	CHICO TOYS	18	CO	76	WFE; Utility; Similar to Lion-Molberg M.F. (130106-007).
☐	130106-012A	50-B	1/43	D	CORGI	50	GB	73	WFE; Industrial; Cab; Yellow and black.
☐	130106-012B	50-B	1/43	D	CORGI	54	GB	74	WFE; Industrial; Same except having a front loader.
☐	130106-012C	50-B	1/43	D	CORGI	54	GB	??	WFE; Same except "Block" decal and orange.
☐	130106-013A	65	1/43	D	CORGI	50	GB	59	WFE; Utility.
☐	130106-013B	65	1/43	D	CORGI	53	GB	59	WFE; Utility; With a front loader.
☐	130106-013C	65	1/43	D	CORGI	57	GB	59	WFE; Utility; With a front fork loader.
☐	130106-014A	165	1/43	D	CORGI	61	GB	67	WFE; Utility; Available with a four-furrow plow.
☐	130106-014B	165	1/43	D	CORGI	66	GB	66	WFE; Utility.
☐	130106-014C	165	1/43	D	CORGI	69	GB	69	WFE; Utility; With front loader.
☐	130106-014D	165	1/43	D	CORGI	73	GB	70	WFE; Utility; With side mounted saw trimmer.
☐	130106-015	165	1/78	D	CORGI (JR)	43	GB	69	WFE; Utility; Cab; With blade; Yellow.
☐	130106-016	65	1/43	D	TRIANG SPOT-ON	137	GB	64	WFE; Utility; Yellow wheels; Three-point-hitch. This is a Massey-Harris-Ferguson 65.

CODE	MODEL	SCALE	MATER.	MANU.	STOCK #	CTRY.	YEAR	REMARKS
☐	130106-017A	65X	1/43	D	JUE		BR	72	WFE; Utility; Excellent detail.
☐	130106-017B	65X	1/43	D	MINIMAC		BR	75	WFE; Utility; Excellent detail; Change in manufacturing ownership.
☐	130106-018A	275	1/43	D	MINIMAC		BR	75	WFE; First version; Has vertical exhaust.
☐	130106-018B	275	1/43	D	MINIMAC		BR	75	WFE; Second version; No vertical exhaust.
☐	130106-019	3366	1/43	D	JUE		BR	72	Industrial crawler with blade.
☐	130106-020A	135 DIESEL	1/38	D	REINDEER	175	ZA	69	WFE; Utility.
☐	130106-020B	(135 DIESEL)	1/38	D	REINDEER		ZA	70	WFE; Utility; Green; Without decals; With four-wheel wagon.
☐	130106-021	175 DIESEL	1/20	D	REINDEER	175	ZA	69	WFE; Utility; "GOODYEAR" tires; Excellent detail; Steerable; Rare.
☐	130106-022A	135	1/32	D	BRITAINS LTD.	9529	GB	69	WFE; Utility; Fiberglass type cab; Some cabs have "glass" windows while others did not.
☐	130106-022B	135 INDUSTRIAL	1/32	D	BRITAINS LTD.	9572	GB	70	WFE; Industrial utility; Cab; Front loader; Yellow and red.
☐	130106-022C	135	1/32	D	BRITAINS LTD.	9520	GB	76	WFE; Utility; No cab.
☐	130106-023A	595	1/32	D	BRITAINS LTD.	9522	GB	76	WFE; Cab.
☐	130106-023B	590	1/32	D	BRITAINS LTD.	9522	GB	80	WFE; Utility; Cab; Very few made with the "590" decals.
☐	130106-023C	595	1/32	D	BRITAINS LTD.	9529	GB	81	WFE; Cab; Dual rear wheels.
☐	130106-024	165	1/43	D	LESNEY	K-3	GB	70	WFE; Utility; Cab; Yellow wheels; With trailer.
☐	130106-025A	165	1/43	D	JOAL	203	E	70	WFE; Utility; Resembles CORGI M.F. (130106-014); "FABRICACION EBRO" on side decals. See also EBRO.
☐	130106-025B	165	1/43	D	JOAL	206	E	72	WFE; Same except with front loader. See also EBRO.
☐	130106-026	165	1/20	PK	SCALE CRAFT	S-523	GB	75	WFE; Cab; Utility; With battery-electric motor; This is a later model than 130106-064.
☐	130106-027	300	1/50	D	NZG	129	D	74	Industrial crawler with blade; Yellow.
☐	130106-028	450-S	1/50	D	NZG	106	D	74	Industrial hydraulic excavator; Yellow.
☐	130106-029A	175 DIESEL	1/16	D	ERTL	175	USA	65	WFE; Decal variations; Metal or plastic wheels; With or without front weight bracket; Some weight brackets are red while others are gray.
☐	130106-029B	3165	1/16	D	ERTL		USA	67	WFE; Industrial tractor with front loader; Yellow.
☐	130106-030	275 DIESEL	1/16	D	ERTL	1103	USA	75	WFE; ROPS; Either narrow or wide front wheels; Decal variations.
☐	130106-031A	1080	1/16	D	ERTL	180	USA	70	WFE; Cab.
☐	130106-031B	1080 V-8	1/16	D	ERTL	180	USA	70	WFE; Cab; This model is mis-labeled as there is no real 1080 with a V-8 engine.
☐	130106-031C	1150	1/16	D	ERTL	179	USA	70	WFE; Cab; This model is mis-labeled as there is no real 1150 with a six cylinder engine.
☐	130106-031D	1150 V-8	1/16	D	ERTL	179	USA	70	WFE; Cab; Dual rear wheels.
☐	130106-032A	1105	1/16	D	ERTL	161	USA	73	WFE; Cab; Red wheels.
☐	130106-032B	1105	1/16	D	ERTL	161	USA	75	WFE; Cab; Gray wheels.
☐	130106-033A	1155	1/64	D	ERTL	1350	USA	73	WFE; Cab; Small wheels.
☐	130106-033B	1155	1/64	D	ERTL	1350	USA	75	WFE; Cab; Large wheels.
☐	130106-034A	1155	1/16	D	ERTL	183	USA	74	WFE; Cab; Red wheels.
☐	130106-034B	1155	1/16	D	ERTL	183	USA	75	WFE; Cab; Gray wheels.
☐	130106-034C	1155	1/16	D	ERTL	183	USA	76	WFE; Cab; Decal variations.
☐	130106-035	1155	1/25	PK	ERTL	8007	USA	75	WFE; Cab.
☐	130106-036	1155	1/25	PK	ERTL	8016	USA	76	WFE; Cab; "SPIRIT OF AMERICA" Commemorative Bicentennial issue.
☐	130106-037A	590	1/16	D	ERTL	1106	USA	77	WFE; Cab; European model; Gray cab.
☐	130106-037B	595	1/16	D	ERTL	1106	USA	77	WFE; Cab; European model; Gray cab; 2600 made.
☐	130106-037C	590	1/16	D	ERTL		USA	80	WFE; ROPS; "TOY FARMER" National Show tractor; The date 11/7/80 is inscribed on the tractor.
☐	130106-038	165	1/16	D	??		RA	65	WFE; Similar to ERTL 165 (130106-029A), but having greater detail including a three-point-hitch; vertical muffler and air cleaner; Rare.
☐	130106-039A	2775	1/64	D	ERTL	1622	USA	78	WFE; Cab.
☐	130106-039B	2800	1/64	D	ERTL	1622	USA	78	WFE; Cab.
☐	130106-040	HANDMAG D600C	1/50	D	CURSOR	1269	D	78	INDUSTRIAL CRAWLER WITH BLADE; CAB.
☐	130106-041	HANDMAG 66C	1/50	D	CURSOR	569	D	78	INDUSTRIAL WHEEL LOADER.
☐	130106-042A	2775	1/20	D	ERTL	1107	USA	79	WFE; Cab with "glass" windows.
☐	130106-042B	2805	1/20	D	ERTL	1108	USA	79	WFE; Cab with "glass" windows; Duals.
☐	130106-043A	1134	1/43	D & P	YAXON	067	I	81	WFE; Front-wheel-assist; Either cab or

CODE	MODEL	SCALE	MATER.	MANU.	STOCK #	CTRY.YEAR	REMARKS
							half cab; Grill and side decal variations; Also available in two-wheel-drive.
130106-043B	1114	1/43	D & P	YAXON	067	I 81	WFE; Front-wheel-assist; Either cab or half cab; Grill and side decal variations; Also available in two-wheel-drive.
130106-044	50B	1/20	C/B	??		GB 80	WFE; Industrial backhoe-loader; Cardboard kit.
130106-045	(135)	1/16	P	BULL		D ??	WFE; Red, green, white and silver.
130106-046	(275)	1/43	D	JOAL	203	E 80	WFE; Lime green; "FABRICACION EBRO" on decals.
130106-047	(1155)	1/70	D	TOMY	F 54	J 81	WFE; Cab; Decalled "MODEL 90 TOUGH".
130106-048	??	1/43	D	YAXON	0501	I 80	WFE; Industrial four-wheel-drive; #0501 with loader; #0502 with loader and backhoe; #503 with v-blade; #504 with blade and backhoe.
130106-049	??	1/87	D	CHARMERZ		HK 80	WFE.
130106-050	EBRO 470	1/43	D	JOAL	206	E 81	WFE; Industrial tractor with front loader.
130106-051	4880	1/64	D	ERTL	1727	USA 82	WFE; Four-wheel-drive; Articulated; Cab; Duals.
130106-052A	4880 COLL. SER.	1/32	D	ERTL	A637	USA 82	WFE; Four-wheel-drive; Articulated; Cab; "COLLECTORS SERIES"; Also a Canadian COLLECTORS SERIES version.
130106-052B	4880	1/32	D	ERTL	1637	USA 82	WFE; Four-wheel-drive; Articulated; Cab.
130106-052C	4900	1/32	D	ERTL	1691	USA 82	WFE; Four-wheel-drive; Articulated; Cab.
130106-053	(165)	8-1/2"	D	DURAVIT		RA 80	WFE.
130106-054A	284-S	1/32	D	SIKU	2550	D 83	WFE.
130106-054B	284-S	1/32	D	SIKU	2570	D 83	WFE; Cab.
130106-054C	284-S	1/32	D	SIKU	3750	D 83	WFE; Front-wheel-assist; Cab; With #2251 side dump wagon.
130106-055	2640	1/32	D	GAMA	2310	D 83	WFE; Front-wheel-assist; Cab.
130106-056	(135)	1/64	D	PLAYART	7591/E	HK 83	WFE; Available with either a roto-tiller or a disk-roller.
130106-057	670	1/20	D	ERTL	1105DO	USA 83	WFE; Cab; Also with Collectors decals.
130106-058A	270	1/16	D	ERTL	1104DO	USA 83	WFE.
130106-058B	270 COLL. SER.	1/16	D	ERTL	1104A	USA 83	WFE; "PHOENIX 83 - A NEW WAY" inscribed on collectors version.
130106-059A	698	1/20	D	ERTL	1102DO	USA 83	WFE; Cab; Dual rear wheels.
130106-059B	698	1/20	D	ERTL	1102DO	USA 83	WFE; Cab; Dual rear wheels; "SPECIAL EDITION" Collectors Series.
130106-060	2680	1/24	PK	HELLER BOBCAT (HUMBROL)	3501	F 83	WFE; Front-wheel-assist; Cab.
130106-061	2680	1/32	D	BRITAINS LTD.	9520	GB 84	WFE; Front-wheel-assist; Cab.
130106-062	(2680)	1/32	D	LONESTAR	1770	GB 84	WFE; Cab; Same castings as Lonestar Case.
130106-063	699	1/64	D	ERTL	1120FO	USA 84	WFE; Cab.
130106-064	165	1/20	PK	SCALE CRAFT		GB ??	WFE; Red and gray; Seat on rear fender; Earlier model than 130106-026; No cab.
130106-065	(165)	1/65	D	PLAYART	7176	HK ??	WFE.
130106-066	(1155)	1/74	D	ZYLMEX	1359	HK ??	WFE; Yellow.
130106-067A	50B	1/35	D	CONRAD	1951	D ??	Industrial loader-backhoe.
130106-067B	500 ELITE	1/35	D	CONRAD	2852	D ??	Industrial loader-backhoe.
130106-068	??	1/87	WM KIT	LANGLEY MINIATURES		GB ??	WFE.
130106-069	55	1/16	CUS.	OLD TIME COLLECTABLES		USA 83	WFE; Very good detail; See also Massey-Harris 55 (130108-043).

130108-000 MASSEY-HARRIS NOW A PART OF MASSEY-FERGUSON LTD.

CODE	MODEL	SCALE	MATER.	MANU.	STOCK #	CTRY.YEAR	REMARKS
130108-001A	(44)	1/43	D	DINKY	27a	GB 48	WFE; Metal wheels/tires; Separate tan driver.
130108-001B	(44)	1/43	D	DINKY	300 or 310	GB 54	WFE; Rubber tires; Separate painted driver; See also Massey-Ferguson (130106-001).
130108-002	(55)	1/80	D	DINKY-DUBLO	069	GB 59	WFE; Utility; Blue.
130108-003	(44)	1/87	D	WARDI BJW	82	GB 50	WFE; With cast-in driver; Fenders over rear wheels.
130108-004	(44)	1/87	D	A.H.I.		HK ??	WFE; With cast-in driver; With gray wheels.
130108-005	(44)	1/87	D	FUN-HO	1	NZ 76	WFE; With cast-in driver; Silver wheels/tires.
130108-006	(44)	3"	D	FUN-HO	309	NZ ??	WFE.
130108-007	(44)	6-1/2"	D	FUN-HO	305	NZ ??	WFE.
130108-008	(44)	1/87	D	LESNEY	4a	GB 54	WFE; With cast in driver; With or without fenders over rear wheels.

	CODE	MODEL	SCALE	MATER.	MANU.	STOCK #	CTRY.	YEAR	REMARKS
☐	130108-009	745D	1/15	D	MOKO-LESNEY	1	GB	51	WFE; Steerable; Excellent detail; Rare.
☐	130108-010A	(745)	1/20	P	RAPHAEL LIPKIN	1091	GB	??	WFE; Blue with red wheels.
☐	130108-010B	745	1/20	P	RAPHAEL LIPKIN		GB	??	WFE; Red with yellow wheels; Promo.
☐	130108-011	(44)	1/43	D	MICRO-MODELS	4322	NZ	54	WFE; Excellent detail; Also available with front loader.
☐	130108-012	(44)	1/20	D	MAJOR-MODELS		NZ	54	WFE; Steerable; Excellent detail; Also available with front loader.
☐	130108-013	(44)	1/20	D	MAJOR-MODELS		NZ	57	Crawler; Based upon castings used for 130108-012.
☐	130108-014	745	1/38	D	P.M.I.	744	ZA	??	WFE; Excellent detail; Also available with implements.
☐	130108-015A	44 STANDARD	1/20	CA	LINCOLN SPECIALITIES	918	CDN	50	WFE; Has screw-in exhaust and air cleaner; Wide fenders over rear wheels.
☐	130108-015B	44 STANDARD	1/20	CA	LINCOLN SPECIALITIES	918	CDN	50	WFE; Same except fenders do not extend over the rear wheels.
☐	130108-016A	44	1/16	D	LINCOLN SPECIALITIES		CDN	??	WFE; With "Dominion Royal" tires; Also available with front loader.
☐	130108-016B	44	1/16	D	LINCOLN SPECIALITIES		CDN	??	WFE; With "Goodyear" tires; Also available with front loader.
☐	130108-016C	44	1/16	D	LINCOLN SPECIALITIES		CDN	??	WFE; With wood wheels and rubber tires, Also available with front loader.
☐	130108-017A	44	1/16	D	SLIK TOYS		USA	??	Row crop; Driver with helmet style hat; Bottom of tractor closed.
☐	130108-017B	44	1/16	CA	SLIK TOYS		USA	??	Row crop; Driver with helmet style hat; Bottom of tractor open. Similar in appearance as 130108-017B.
☐	130108-018	44	1/16	CA	THE KING COMPANY		USA	??	Row crop; Separately cast driver with "hard hat"; Driver mounting hole at front of platform.
☐	130108-019	44	1/16	CA	THE KING COMPANY		USA	??	Row crop; Separately cast driver with "baseball" type cap; Driver mounting hole at rear of platform.
☐	130108-020A	44	1/16	D	RUEHL PRODUCTS	M44	USA	54	Row crop; Steerable; Metal rear wheels; Excellent detail.
☐	130108-020B	44	1/16	D	RUEHL PRODUCTS	M44	USA	54	Row crop; Steerable; Plastic rear wheels; Excellent detail.
☐	130108-021	(44)	1/87	D	GITANES	1	F	??	WFE; Gray metal wheels/tires or rubber tires; Available with a two wheel trailer.
☐	130108-022A	CHALLANGER	1/12	CA	ROBERT GRAY	NONE	USA	71	Row crop; Represents a 1940 model.
☐	130108-022B	CHALLANGER	1/12	K	ROBERT GRAY	NONE	USA	75	Row crop; Same patterns used but made of korloy instead of cast aluminum.
☐	130108-022C	CHALLANGER	1/12	K	OLD TIME COLLECTABLES	NONE	USA	83	Row crop; Same patterns used but different manufacturer.
☐	130108-023A	(745)	1/20	D	LINCOLN MICRO MODELS		HK	50	WFE; Steerable; Red with yellow wheels and black and silver engine; "Empire made" cast on hood side; With or without muffler and air cleaner stacks; Also available with a front loader.
☐	130108-023B	(745)	1/20	D	LINCOLN MICRO MODELS		HK	50	Crawler with blade; Based upon the same castings as used for the (745) wheel tractor.
☐	130108-024	(55)	6-1/2"	P	PLASTICUM			??	WFE; Clockwork; Steerable; Either red with yellow trim or yellow with red trim.
☐	130108-025	(44)	6"	D	??		NZ	??	WFE.
☐	130108-026	??		D	JOAL	210	E	??	Crawler with blade.
☐	130108-027	(44)	4"	P	MITOPLAST	320	BR	82	WFE; Red or blue.
☐	130108-028	(44)	8"	P	MITOPLAST	320	BR	83	WFE; Red or blue.
☐	130108-029A	744	1/16	CUS.	MARBIL	NONE	GB	82	Row crop; Very limited production.
☐	130108-029B	744	1/16	CUS.	MARBIL	NONE	GB	82	WFE; Very limited production.
☐	130108-029C	744	1/16	CUS.	MARBIL	NONE	GB	83	WFE; With halftracks; Very limited production.
☐	130108-030	20K	1/28	D & T	BP		DK	50	WFE; Also available with three-bottom plow, trailer, roller, seeder and reaper.
☐	130108-031	PONY 820	1/20	P	??		F	69	WFE; Came with a plow; Red; Rare.
☐	130108-032	??	1/43	D	??		DK	50	WFE.
☐	130108-033	(15-22)	1/16	CA	SCALE MODELS	NO. 11	USA	83	WFE; Four-wheel-drive; "Steel" wheels; No. 11 in the JLE Collector Series; 5000 made.
☐	130108-034	(44)	1/30	P	TUDOR ROSE		GB	60	Row crop. MISC 63.
☐	130108-035	(44)	1/30	P	POLISTIL		I	60	Row crop; Copy of Tudor Rose
☐	130108-036	(44)	1/87	D	JADALI		F	??	WFE.
☐	130108-037	(44)	1/87	D	MT		J	??	WFE.
☐	130108-038	(44)		P	??			??	WFE.
☐	130108-039	44	1/20	P	KARKURO	596	D	??	WFE.
☐	130108-040	44	1/30	P	KARKURO	601	D	??	WFE.
☐	130108-041	44	1/15	P	KARKURO	608	D	??	WFE.

	CODE	MODEL	SCALE	MATER.	MANU.	STOCK #	CTRY.	YEAR	REMARKS
☐	130108-042	55	6-1/2"	P	RELIABLE		CN	??	WFE; Similar to Plasticum Massey-Harris 130108-024.
☐	130108-043	55	1/16	CUS.	OLD TIME COLLECTABLES	NONE	USA	83	WFE; Very good detail; See also Massey-Ferguson 55 (030106-069).
☐	130108-044	44	1/16	CA	JIM HOSCH	NONE	USA	80	Row crop; Re-issue of Slik 44 (130108-017B).
☐	130108-045	(44)	4-1/2"	D	EMPIRE MADE			??	WFE; Red with white wheels.
☐	130108-046	(44)	2"	P	EMPIRE MADE			??	WFE; With driver; Color variations.
☐	130108-047	(44)		P	??			??	WFE.
☐	130108-048	CHALLANGER	1/43	D	ERTL	2511	USA	85	Row crop; Represents a 1936 model.

130114-000   MAN

	CODE	MODEL	SCALE	MATER.	MANU.	STOCK #	CTRY.	YEAR	REMARKS
☐	130114-001	??	1/32	P	HERBART		D	60	WFE; Gray; Variations exist.

130115-000   MANITOU

	CODE	MODEL	SCALE	MATER.	MANU.	STOCK #	CTRY.	YEAR	REMARKS
☐	130115-001	FORK LIFT	1/25	P	BOURBON		F	83	Industrial fork lift; Red and white; Variations exist.

130502-000   MERCEDES-BENZ

	CODE	MODEL	SCALE	MATER.	MANU.	STOCK #	CTRY.	YEAR	REMARKS
☐	130502-001A	MB TRAC	1/90	P	WIKING	V-385	D	74	WFE; Four-wheel-drive; Cab.
☐	130502-001B	MB TRAC	1/90	P	WIKING		D	??	WFE; Four-wheel-drive; Cab; With grit spreader.
☐	130502-002A	UNIMOG	1/32	D	BRITAINS LTD.	9569	GB	76	Four-wheel-drive; Used either as a truck or a tractor; Green and yellow.
☐	130502-002B	UNIMOG	1/32	D & P	BRITAINS LTD.	9595	GB	83	Four-wheel-drive; With snow plow and grit spreader; Yellow.
☐	130502-002C	UNIMOG	1/32	D	BRITAINS LTD.	9813	GB	84	Four-wheel-drive; Same except all yellow; "Autoway Series" Also with snow plow and grit spreader (9884) in the Autoway Series.
☐	130502-003	MB TRAC 1300		D	GESCHA	3043	D	77	WFE; Four-wheel-drive; Cab; With or without top exhaust.
☐	130502-004	MB TRAC 1300	1/20	D & P	GAMA	433	D	77	WFE; Four-wheel-drive; Cab.
☐	130502-005	1300 MB	1/20	P	BRUDER		D	82	WFE; Four-wheel-drive; Cab.
☐	130502-006A	1500 MB	1/32	D & P	BRITAINS LTD.	9525	GB	82	WFE; Four-wheel-drive; Yellow and black; Cab; Green.
☐	130502-006B	1500 MB	1/32	D & P	BRITAINS LTD.	9597	GB	84	WFE; Four-wheel-drive; With tipping hopper; Red and white; Cab.
☐	130502-007	UNIMOG	29 CM	P	MEHAND-TEKNICA		YUG	82	Four-wheel-drive.
☐	130502-008	UNIMOG (411)	1/25	T	TCO	U34/411		55	Four-wheel-drive; Green.
☐	130502-009	MB TRAC	1/87	P	BRUDER		D	74	WFE; Four-wheel-drive; Available with a variety of implements.
☐	130502-010	MB TRAC 800	1/32	D	SIKU	2852	D	84	WFE; Four-wheel-drive; Three-point-hitch; Cab.
☐	130502-011	UNIMOG	1/87	P	WIKING	37A	D	55	Four-wheel-drive; The first Unimog; Also with snow plow.
☐	130502-012	UNIMOG 406	1/87	P	WIKING	37G	D	69	Four-wheel-drive.
☐	130502-013	UNIMOG 411	1/87	P	WIKING	37E	D	59	Four-wheel-drive.
☐	130502-014	UNIMOG 406	1/87	P	WIKING	64K	D	66	Four-wheel-drive.
☐	130502-015	UNIMOG 1700L	1/87	P	WIKING	375	D	78	Four-wheel-drive.
☐	130502-016A	UNIMOG 406	1/50	D	CORGI	406	GB	68	Four-wheel-drive; Rubber tires; Rear view mirror.
☐	130502-016B	UNIMOG 406	1/50	D	CORGI	406	GB	76	Four-wheel-drive; Plastic tires; No rear view mirror.
☐	130502-017	UNIMOG 406	1/68	D	LESNEY	49	GB	67	Four-wheel-drive.
☐	130502-018	UNIMOG 406	1/50	D	SIKU	1620	D	??	Four-wheel-drive.
☐	130502-019	UNIMOG 406	1/50	D	MARKLIN	1830	D	67	Four-wheel-drive; Also available with industrial implements.
☐	130502-020	UNIMOG 406	1/71	D	TOMICA	F 41	J	80	Four-wheel-drive.
☐	130502-021	UNIMOG 411	1/66	D	POLISTIL	J26	I	67	Four-wheel-drive.
☐	130502-022	UNIMOG 411	1/160	P	ARNOLD	6608	D	70	Four-wheel-drive.
☐	130502-023	UNIMOG 425	1/43	D	LESNEY	K 37	GB	75	Four-wheel-drive.
☐	130502-024	UNIMOG 425	1/60	D	CURSOR	974	D	??	Four-wheel-drive.
☐	130502-025	UNIMOG 425	1/87	P	CURSOR		D	??	Four-wheel-drive.
☐	130502-026	UNIMOG 406	1/25	P	CURSOR		D	??	Four-wheel-drive.
☐	130502-027	UNIMOG 425	1/30	D	GAMA	427	D	80	Four-wheel-drive.
☐	130502-028	UNIMOG 425	1/82	D	MAJORETTE	415	D	80	Four-wheel-drive; Front fork lift.
☐	130502-029	UNIMOG 1300	1/27	D	SIKU	4510	D	82	Four-wheel-drive.
☐	130502-030	UNIMOG 1300	1/87	P	ROCO	1502	A	82	Four-wheel-drive.
☐	130502-031	UNIMOG 406	1/18	D	MS TOY		D	??	Four-wheel-drive.
☐	130502-032	UNIMOG 411	1/87	P	EKO		E	??	Four-wheel-drive.
☐	130502-033	UNIMOG 425	1/80	D	LESNEY	MB 48	GB	83	Four-wheel-drive with snow plow; Yellow.
☐	130502-034	UNIMOG 1300	1/50	D	SIKU		D	84	Four-wheel-drive.
☐	130502-035	UNIMOG 406	1/90	P	BRUDER		D	??	Four-wheel-drive.
☐	130502-036	MB TRAC	1/87	P	BRUDER		D	82	WFE; Four-wheel-drive.
☐	130502-037	UNIMOG 411	1/60	P	SIKU	V 104	D	60	Four-wheel-drive.

130509-000   MEILI

	CODE	MODEL	SCALE	MATER.	MANU.	STOCK #	CTRY.	YEAR	REMARKS
☐	130509-001	AGROMOBILE	1/60	P	SIKU	V 173	D	??	Agricultural transporter; Green.

130518-000   MERLIN

	CODE	MODEL	SCALE	MATER.	MANU.	STOCK #	CTRY.	YEAR	REMARKS
☐	130518-001	6CV 1902	1/43	CUS.	PHANTOM MODELS	A02	F	??	Small portable gasoline engine; Represents a 1912 model.

130903-000   MICHIGAN   ALSO CLARK-MICHIGAN & CLARK

	CODE	MODEL	SCALE	MATER.	MANU.	STOCK #	CTRY.	YEAR	REMARKS
☐	130903-001	CLARK MICHIGAN 175 C	1/35	D	CONRAD	2932	D	83	Industrial loader.

	CODE	MODEL	SCALE	MATER.	MANU.	STOCK #	CTRY.	YEAR	REMARKS
☐	130903-002	MICHIGAN 180 III	1/43	D	DINKY	976	GB	68	Industrial loader.
☐	130903-003	MICHIGAN 75 B LOADER	1/20	D	GESCHA	501	D	??	Industrial loader; Excellent detail.
☐	130903-004	CLARK 125 B LOADER	1/50	D	CONRAD	2884	D	??	Industrial loader.

131300-000 MINNEAPOLIS-MOLINE NOW A PART OF WHITE FARM EQUIPMENT COMPANY

	CODE	MODEL	SCALE	MATER.	MANU.	STOCK #	CTRY.	YEAR	REMARKS
☐	131300-001	(STANDARD U)	1/12	W	??		USA	44	WFE; Very crude.
☐	131300-002	(ROW CROP Z)	1/12	W	WERNER WOOD & PLASTIC CO.	NONE	USA	44	Row crop; Steerable; Very crude.
☐	131300-003A	Z	1/32	R	AUBURN RUBBER		USA	38	Row crop; Molded-in driver; Variety of colors.
☐	131300-003B	Z	1/32	R	AUBURN RUBBER		USA	38	Row crop; Molded-in driver; Yellow with white wheels; Promo.
☐	131300-004A	(R)	1/16	R	AUBURN RUBBER		USA	51	Row crop; Molded-in driver; Variety of colors; Driver's head turned to one side; Baseball style cap.
☐	131300-004B	(R)	1/16	R	AUBURN RUBBER		USA	45	Row crop; Molded in driver; Variety of colors; Driver's head looking straight foreward; Full brim hat.
☐	131300-005	(UB)	1/16	D	SLIK TOYS	9853	USA	56	Row crop; Steerable.
☐	131300-006	(R)	1/32	D	SLIK TOYS	9816	USA	??	Row crop; Cast-in driver; "M-M" cast on sides; Red, yellow or green.
☐	131300-007	(4 STAR)	1/32	D	SLIK TOYS	9871	USA	??	WFE; Straight vertical bars on grill.
☐	131300-008	445	1/32	D	SLIK TOYS	9871	USA	??	WFE; Cross bars on grill; Lights on sides of grill.
☐	131300-009	(R)	1/32	D	LINCOLN SPECIALITIES		CDN	??	Row crop; Cast-in driver; Variety of colors and wheel/tire styles; Similar to Slik Toys (R) 131300-006.
☐	131300-010A	(R)	1/16	D	SLIK TOYS		USA	50	Row crop; Cast-in driver; Driver has flat top hat. This original version is very scarce.
☐	131300-010B	(R)	1/16	D	HOSCH		USA	81	Row crop; Same molds used to recast this version; M-M letters raised rather than decals like the original.
☐	131300-011A	(602)	1/25	D	ERTL	15	USA	63	Row crop; Plastic wheels/tires; Yellow over bronze; Non-steerable.
☐	131300-011B	(602)	1/25	D	ERTL	15	USA	65	Row crop; Plastic wheels; All yellow; Non-steerable.
☐	131300-012A	(602 LPG)	1/25	D	ERTL	15	USA	63	Row crop; Plastic wheels/tires; Yellow over bronze; Pressure type fuel tank.
☐	131300-012B	(602 LPG)	1/25	D	ERTL	15	USA	65	Row crop; Plastic wheels/tires; All yellow; Pressure type fuel tank.
☐	131300-012C	THERMOGAS (602)	1/25	D	ERTL	15	USA	65	Row crop; Plastic wheels/tires; Yellow; This model was a promo for the Thermogas Company.
☐	131300-013	(ROW CROP LPG)	1/25	D	ERTL	15	USA	67	Row crop; Steerable; Rubber wheels/tires; Yellow; Pressure type fuel tank.
☐	131300-014A	G-1000 VISTA	1/16	D	ERTL	17	USA	68	WFE; Yellow wheels.
☐	131300-014B	(G-1000 VISTA)	1/16	D	ERTL	17	USA	72	WFE; White wheels; No number on decals.
☐	131300-015A	MIGHTY MINNIE	1/16	D	ERTL	2702	USA	74	WFE; Modified G-1000 Pulling tractor; Chrome engine; Color and wheel variations exist.
☐	131300-015B	MIGHTY MINNIE	1/16	D	ERTL	2702	USA	75	WFE; Same except has black engine.
☐	131300-016	G-1355	1/16	D	ERTL	19	USA	74	WFE; Dual rear wheels; Flotation front tires; ROPS; See also Oliver 1855 151209-017K.
☐	131300-017	J	1/16	D	SCALE MODELS	NO. 5	USA	81	Row crop; Represents a 1935 model; No 5 in the JLE Collector Series; "Steel" wheels; 3000 made.
☐	131300-018	R	1/16	CUS.	DENNIS PARKER	NONE	USA	79	Row crop; Represents a 1951 model; 140 made.
☐	131300-019	25 HP POWER UNIT	1/16	W	PLOW BOY TOYS	NONE	USA	81	Represents a 1954 power unit such as those used to power sawmills and irrigation pumps.
☐	131300-020A	A4T-1400	1/16	CUS.	DAVID SHARP	NONE	USA	82	WFE; Four-wheel-drive; Without cab; Duals Articulated.
☐	131300-020B	A4T1600	1/16	CUS.	DAVID SHARP	NONE	USA	82	WFE; Four-wheel-drive; Cab; Articulated; Duals.
☐	131300-021	G-1350	1/16	CUS.	DAVID SHARP	NONE	USA	81	WFE; Red, white and blue "Heritage" tractor or all yellow.
☐	131300-022	UB	1/16	CUS.	DENNIS PARKER	NONE	USA	83	WFE; Based upon the Slik (UB) 151209-005.
☐	131300-023A	COMFORTRACTOR	1/16	D	SCALE MODELS		USA	84	WFE; Cab; Represents a 1938 UDLX; Commemorates "THE PRAIRIE GOLD RUSH, Greenville, Ohio July 4, 1984"; 1200 made.
☐	131300-023B	COMFORTRACTOR	1/16	D	SCALE MODELS		USA	84	WFE; Cab; Same except wihtout commemorative inscription; JLE Antique Series; 5000 made.

	CODE	MODEL	SCALE	MATER.	MANU.	STOCK #	CTRY.	YEAR	REMARKS
☐	131300-024	GTB	1/16	CA	WALLY HOOKER	NONE	USA	84	WFE; Represents a 1955 model; Limited production.
☐	131300-025	G-950	1/16	CUS.	DAVID SHARP	NONE	USA	81	WFE; Cab; Yellow.
☐	131300-026	G-1050	1/16	CUS.	DAVID SHARP	NONE	USA	81	WFE; Cab; Yellow.
☐	131300-030	??	1/43	D	CHARBENS		GB	??	WFE; Clockwork.
☐	131300-031	TWIN CITY	1/43	BRASS	PHANTOM MODELS	A03	F	83	WFE; Very limited production.
☐	131300-032A	Z	1/16	CA	A T & T COLLECTABLES		USA	83	Row crop; Spoke wheels; Rubber; Fenders.
☐	131300-032B	Z	1/16	CA	A T & T COLLECTABLES	NONE	USA	83	Row crop; Solid wheels; Rubber; Fenders.
☐	131300-033A	Z	1/16	CA	A T & T COLLECTABLES	NONE	USA	83	Row crop; Cast in driver; Flat fenders.
☐	131300-033B	Z	1/16	CA	A T & T COLLECTABLES	NONE	USA	83	Row crop; Cast in driver; Round fenders.
☐	131300-034	R	1/16	CA	A T & T COLLECTABLES	NONE	USA	83	Row crop.
☐	131300-035A	U	1/16	CA	A T & T COLLECTABLES	NONE	USA	83	WFE.
☐	131300-035B	U	1/16	CA	A T & T COLLECTABLES	NONE	USA	83	Row crop.
☐	131300-036	R	1/16	CA	A T & T COLLECTABLES	NONE	USA	84	Row crop; Cast in driver; Cab.
☐	131300-037	UB LPG	1/12	WOOD	GUBBELS	NONE	USA	80	Row crop; LP fuel tank.

131514-000 MONARCH

	CODE	MODEL	SCALE	MATER.	MANU.	STOCK #	CTRY.	YEAR	REMARKS
☐	131514-001	??	SMALL	CI	HUBLEY		USA	??	Crawler.
☐	131514-002	??	MED	CI	HUBLEY		USA	??	Crawler.
☐	131514-003	??	LARGE	CI	HUBLEY		USA	??	Crawler.
☐	131514-004	??	5-1/2"	CA	OTT & PTW	NONE	USA	??	Crawler; Old Time Toys models is orange while Pioneer Tractor Works model is olive drab green.

131913-000 M. S. & M. CO.   MINNEAPOLIS STEEL AND MACHINERY CO.

	CODE	MODEL	SCALE	MATER.	MANU.	STOCK #	CTRY.	YEAR	REMARKS
☐	131913-001	TWIN CITY 60/90	1/43	CUS.	PHANTOM MODELS	A03	F	??	WFE; See also Twin City.

132108-000 MUIR-HILL

	CODE	MODEL	SCALE	MATER.	MANU.	STOCK #	CTRY.	YEAR	REMARKS
☐	132108-001A	2WL	1/43	D	DINKY	437	GB	62	WFE; Industrial tractor with front loader; Red or yellow.
☐	132108-001B	2WL	1/43	D	DINKY	967	GB	73	WFE; Industrial tractor with front loader and backhoe; Yellow.
☐	132108-002A	161	1/50	D	LESNEY	K-5	GB	72	Four-wheel-drive; Industrial tractor with blade and tandem trailer.
☐	132108-002B	161 MH6	1/50	D	LESNEY	K-25	GB	??	Industrial four-wheel tractor backhoe and blade.
☐	132108-002C	161	1/50	D	LESNEY	1	GB	84	Four-wheel-drive industrial tractor with backhoe; Only sold with articulated truck.

140800-000 NEW HOLLAND

	CODE	MODEL	SCALE	MATER.	MANU.	STOCK #	CTRY.	YEAR	REMARKS
☐	140800-001	1/2 HP	1/16	CUS.	ALVIN EBERSOL	NONE	USA	77	Represents an antique gasoline engine.
☐	140800-002	GAS ENGINE	1/16	CA	SCALE MODELS	NONE	USA	82	Represents an antique gasoline engine; On wheels; 5000 made; JLE Specialities Series.

140900-000 NEW IDEA

	CODE	MODEL	SCALE	MATER.	MANU.	STOCK #	CTRY.	YEAR	REMARKS
☐	140900-001A	GAS ENGINE	1/16	CA	ROBERT GRAY	NONE	USA	81	Represents an antique gasoline engine.
☐	140900-001B	GAS ENGINE	1/16	CA	ROBERT GRAY	NONE	USA	81	Same except on wheels.

141500-000 NORM

	CODE	MODEL	SCALE	MATER.	MANU.	STOCK #	CTRY.	YEAR	REMARKS
☐	141500-001	NORMAL	1/90	P	WIKING	38	D	??	WFE; Red or green.

141518-000 NORMAG ZORGE

	CODE	MODEL	SCALE	MATER.	MANU.	STOCK #	CTRY.	YEAR	REMARKS
☐	141518-001	ZORGE	1/70	P	CURSOR		D	58	WFE; Available with mower and plow.

142106-000 NUFFIELD

	CODE	MODEL	SCALE	MATER.	MANU.	STOCK #	CTRY.	YEAR	REMARKS
☐	142106-001	UNIVERSAL (M-IV)	1/16	D	DENZIL SKINNER		GB	54	WFE; Steerable; Red; Excellent detail; Rare.
☐	142106-002	(UNIVERSAL)	1/15	P	RAPHAEL LIPKIN	1078	GB	63	WFE.
☐	142106-003	(UNIVERSAL)	1/16	P	MINIC	2	GB	54	WFE; With driver; Clockwork mechanism; Rear tires have a series of holes through which a key can be inserted to wind the clockwork located in the rear of the tractor.
☐	142106-004A	(UNIVERSAL)	1/48	D	DENZIL SKINNER		GB	54	Row crop; Excellent detail; Rare.
☐	142106-004B	(UNIVERSAL)	1/48	D	DENZIL SKINNER		GB	54	WFE; Steerable; Excellent detail; Red. Rare.

151209-000 OLIVER   NOW A PART OF WHITE FARM EQUIPMENT

	CODE	MODEL	SCALE	MATER.	MANU.	STOCK #	CTRY.	YEAR	REMARKS
☐	151209-001	(70)	1/25	CI	ARCADE	359	USA	36	Row crop; Cast-in driver; Red; Rubber wheels/tires.
☐	151209-002	70	1/16	CI	ARCADE	3560	USA	36	Row crop; Separately cast, plated driver; Red or green.
☐	151209-003	70 ORCHARD	1/25	CI	HUBLEY	612	USA	38	WFE; Separately cast driver; Fenders over rear wheels; Wood or rubber wheels/tires.

	CODE	MODEL	SCALE	MATER.	MANU.	STOCK #	CTRY.	YEAR	REMARKS
☐	151209-004A	70	1/16	CA	SLIK TOYS		USA	47	Row crop; Similar to Arcade 70 (151209-002).
☐	151209-004B	70	1/16	CA	HOSCH		USA	81	Row crop; Re-issue of Slik Toys 70 (151209-004A).
☐	151209-005A	77	1/16	D	SLIK TOYS		USA	48	Row crop; Separately cast driver.
☐	151209-005B	77 ROW CROP	1/16	D	SLIK TOYS		USA	50	Row crop; Separately cast driver.
☐	151209-006A	77 R.C. DIESEL POWER	1/16	D	SLIK TOYS		USA	52	Row crop; No driver; Steerable; Closed engine; Red wheels.
☐	151209-006B	77 R.C. DIESEL POWER	1/16	D	SLIK TOYS		USA	54	Row crop; No driver; Steerable; Open engine; Green wheels.
☐	151209-007	SUPER 77	1/16	D	SLIK TOYS		USA	54	Row crop; No driver; Open engine; Steerable; Green wheels.
☐	151209-008	SUPER 55	1/12	D	SLIK TOYS		USA	55	WFE; Utility; Non-steering; Three-point-hitch; Also available was a three-point-hitch plow; Tire variations exist.
☐	151209-009	(880)	1/32	CA	SLIK TOYS	9876	USA	60	Row crop; Non-steering.
☐	151209-010	880	1/16	CA	SLIK TOYS		USA	58	Row crop; Non-steering.
☐	151209-011A	OC-6	1/16	D	SLIK TOYS	9851	USA	??	Crawler; Wide track row crop style; Separate cast driver; Yellow.
☐	151209-011B	OC-6	1/16	D	MAASDAM	NONE	USA	83	Crawler; Reissue using original Slik Toys dies.
☐	151209-012	70	1/20	R	AUBURN RUBBER	543	USA	??	Row crop; Molded in driver; Red.
☐	151209-013	77 STANDARD	1/16	CA	LINCOLN SPECIALITIES		CDN	50	WFE; Standard style tractor; No driver.
☐	151209-014	70 ORCHARD	1/25	CA	FUN-HO	81	NZ	??	WFE; Reproduction of Hubley 70 Orchard tractor (151209-003).
☐	151209-015A	HART-PARR (28-44)	1/12	CA or K	ROBERT GRAY	NONE	USA	71	WFE; Represents a 1930's standard style tractor; Separately cast driver.
☐	151209-015B	HART-PARR (28-44)	1/12	K	ROBERT GRAY	NONE	USA	79	WFE; Represents a 1930's standard style tractor; Rubber tires; Robert Gray's 10th Anniversary Issue; Gold colored driver.
☐	151209-016A	70 RC	1/16	CA or K	ROBERT GRAY	NONE	USA	76	Row crop; "Steel" wheels; Separately cast driver.
☐	151209-016B	70 STANDARD	1/16	K	ROBERT GRAY	NONE	USA	76	WFE; Standard style tractor; "Steel" wheels; Driver.
☐	151209-016C	70 STANDARD	1/16	K	ROBERT GRAY	NONE	USA	71	WFE; Gold colored driver; Robert Gray's 10th Anniversary Issue; Rubber tires.
☐	151209-017A	1800 (SERIES A)	1/16	D	ERTL	604	USA	63	Row crop; Steerable; "Checkerboard" decals with red keystone.
☐	151209-017B	1800 (SERIES B)	1/16	D	ERTL	604	USA	64	Row crop; "Checkerboard" decals.
☐	151209-017C	1800 (SERIES C)	1/16	D	ERTL	604	USA	64	Row crop; Large 1800 numbers and red trim on decals.
☐	151209-017D	1800 FWA	1/16	D	ERTL	606	USA	63	WFE; Front-wheel-assist.
☐	151209-017E	1850	1/16	D	ERTL	604	USA	65	Row crop.
☐	151209-017F	1850 FWA	1/16	D	ERTL	606	USA	65	WFE; Front-wheel-assist.
☐	151209-017G	1850	1/16	D	ERTL	604	USA	68	Row crop; Without fenders; Wheels and grill are plastic instead of diecast metal like the earlier models.
☐	151209-017H	WHITE 1855	1/16	D	ERTL	604	USA	70	Row crop; No fenders.
☐	151209-017I	WHITE 1855 FWA	1/16	D	ERTL	606	USA	70	WFE; Front-wheel-assist.
☐	151209-017J	WHITE 1855	1/16	D	ERTL	609	USA	75	WFE; Flotation front tires.
☐	151209-017K	1855	1/16	D	ERTL	610	USA	74	WFE; Dual rear wheels; ROPS; Flotation front tires.
☐	151209-017L	1855	1/16	D	ERTL	604	USA	75	Row crop; No fenders.
☐	151209-018	(1955)	1/43	D	TOOTSIETOY	1435	USA	70	WFE; Came with tandem manure spreader.
☐	151209-019	(70)	1/16	CA	E.R. ROACH		USA	??	Row crop; With cast-in driver.
☐	151209-020	ROW CROP 80	1/16	CA	SCALE MODELS	NO. 3	USA	80	Row crop; "Steel" wheels; Represents a 1937 model; No. 3 in the JLE Collector Series; 3000 made.
☐	151209-021A	2655	1/16	CUS.	DAVID SHARP	NONE	USA	80	WFE; Four-wheel-drive; Cab.
☐	151209-021B	2455	1/16	CUS.	DAVID SHARP	NONE	USA	80	WFE; Four-wheel-drive; No cab.
☐	151209-022	2155	1/16	CUS.	DAVID SHARP	NONE	USA	80	WFE.
☐	151209-023	70	1/16	D	SCALE MODELS		USA	83	Row crop; "Steel" wheels; Represents a 1938 model; Serial numbered; 5000 made; See also Cockshutt 70 (031503-007); Antique Series.
☐	151209-024A	550	1/16	P	WELDON YODER	NONE	USA	84	WFE; Utility; Steerable; Represents a 1961 model; Slotted style grill; Serial numbered; 3000 made; Oscillating front axle; Swinging drawbal.
☐	151209-024B	550	1/16	P	WELDON YODER	NONE	USA	84	WFE; Utility; Steerable; Represents a 1963 model; Box style grill; Serial numbered; See also Cockshutt 550 (031503-008); 3000 made.
☐	151209-025	770	1/16	CA	BEN SIEGEL	NONE	USA	84	WFE; Represents a 1958 model.
☐	151209-026A	770	1/16	CUS.	LYLE DINGMAN	NONE	USA	84	Row crop; Based on a Slik Oliver 880 (151209-009) castings.

CODE	MODEL	SCALE	MATER.	MANU.	STOCK #	CTRY.	YEAR	REMARKS
☐ 151209-026B	880	1/16	CUS.	LYLE DINGMAN	NONE	USA	84	WFE; Based on Slik Oliver 880 (151209-009) castings.
☐ 151209-027	88 ROW CROP	1/16	RESIN	GARY ANDERSON	NONE	USA	84	WFE.
☐ 151209-028	(LGT)	4"	SS	SLIK		USA	??	Lawn and garden tractor; With trailer.
☐ 151209-029	(70)	1/16	P	??		USA	??	Row crop.
☐ 151209-030	70		W	STURDY STUFF TOYS	NONE	USA	??	Row crop; Made in Decauter, Michigan.

151616-000 OPPERMAN

CODE	MODEL	SCALE	MATER.	MANU.	STOCK #	CTRY.	YEAR	REMARKS
☐ 151616-001	MOTOCART	1/40	D	DINKY	27G	GB	??	Transporter; Three wheels.

152015-000 ODERO TERNI ORLANDO (OTO)   **ODERO TERNI ORLANDO-LA SPEZIA**

CODE	MODEL	SCALE	MATER.	MANU.	STOCK #	CTRY.	YEAR	REMARKS
☐ 152015-001	C 25 R	1/20	CA	MACHPI- BOLOGNA		I	55	WFE; Red; Rare; Only 9000 total wheel tractor and crawler models made; The real tractor has a one cylinder horizontal diesel four stroke engine.
☐ 152015-002	C 25 C	1/20	CA	MACHPI- BOLOGNA		I	56	Crawler; Metal tracks; Red; Rare; Real tractor has one cylinder horizontal four stroke engine.

152301-000 OWATONNA

CODE	MODEL	SCALE	MATER.	MANU.	STOCK #	CTRY.	YEAR	REMARKS
☐ 152301-001	MUSTANG	1/16	D	ERTL	725	USA	77	Four-wheel skid-steer loader; ROPS.
☐ 152301-002A	MUSTANG	1/16	D	ERTL	355	USA	81	Four-wheel skid-steer loader; ROPS.
☐ 152301-002B	MUSTANG COLL. SER.	1/16	D	ERTL	355	USA	81	Four-wheel skid steer loader; ROPS; Serial numbered COLLECTORS version.

160114-000 PANHARD

CODE	MODEL	SCALE	MATER.	MANU.	STOCK #	CTRY.	YEAR	REMARKS
☐ 160114-001	??	1/43	L	AR		F	??	Crawler; Antique style.

160518-000 PERPLEX   **MOVE TO MISC SECTION**

CODE	MODEL	SCALE	MATER.	MANU.	STOCK #	CTRY.	YEAR	REMARKS
☐ 160518-001	7300	1/16	T	ARNOLD		G	??	WFE; Steerable; Battery-electric motor.
☐ 160518-002	7300	1/16	T	ARNOLD			??	Crawler; Battery electric.

160719-000 P.G.S.   **PIGIESSE**

CODE	MODEL	SCALE	MATER.	MANU.	STOCK #	CTRY.	YEAR	REMARKS
☐ 160719-001	420 ROMA	1/9	PK	PROTAR		I	70	WFE; Four-wheel-drive; Articulated; Hood lifts.
☐ 160719-002	ROMA 38	18"	P	MEHAND TEHNIKA	B-470	JUG	84	WFE; Four-wheel-drive; Articulated; Steerable; Red-yellow-green.

161518-000 PORSCHE

CODE	MODEL	SCALE	MATER.	MANU.	STOCK #	CTRY.	YEAR	REMARKS
☐ 161518-001	??	1/43	P & T	GAMA		D	??	WFE; Clockwork; Came with a four wheel wagon.
☐ 161518-002A	DIESEL-T	1/60	D	SIKU	218	D	64	WFE; Red or blue.
☐ 161518-002B	DIESEL-T	1/60	D	SIKU	254	D	64	WFE; Came with a two wheel trailer.
☐ 161518-003	T	1/90	P	WIKING	38P or 380	D	58	WFE; Orange or red.
☐ 161518-004	ERROR							
☐ 161518-005	?? DIESEL	12"	P	??			??	WFE.
☐ 161518-006	??	6-1/4"	VINYL	STEHO		D	76	WFE; With four wheel wagon.
☐ 161518-007	DIESEL STANDARD	5"	P	CURSOR		D	??	WFE.
☐ 161518-008	??	10-1/2"	P	??		D	??	WFE; Two seat model; Orange, green, black and silver.
☐ 161518-009	DIESEL	7-1/2"	P	MS - MICHAEL SEIDEL	1774-2	D	??	WFE.
☐ 161518-010	STANDARD	1/32	P	HERBART		D	60	WFE; Red and green.
☐ 161518-011	ALLGAIER POR. A-133	1/32		MS		D	56	WFE; Orange; Clockwork.
☐ 161518-012	T	1/12	P	MS		D	60	WFE; With or without front loader.

180514-000 RENAULT

CODE	MODEL	SCALE	MATER.	MANU.	STOCK #	CTRY.	YEAR	REMARKS
☐ 180514-001	R 30-40	1/32	D	CIJ	3-33	F	??	WFE; Orange.
☐ 180514-002	E-30	1/32	D	CIJ	3-33	F	59	WFE; Orange.
☐ 180514-003	E-30	1/32	D	CIJ	3-33	F	59	WFE; Red; Different grill.
☐ 180514-004A	R 30-40	1/16	T	CIJ	8-52	F	??	WFE; Clockwork; With or without hay tedder.
☐ 180514-004B	R 30-40	1/16	T	CIJ	3-34	F	??	WFE; Red.
☐ 180514-005	??	1/25	P	CURSOR		D	??	WFE.
☐ 180514-006A	R-86	1/43	P	NOREV	117	F	??	WFE; With rear light; Orange, dark green or blue.
☐ 180514-006B	R-86	1/43	P	NOREV	117	F	??	WFE; Without rear light; Yellow or pale green.
☐ 180514-006C	R-86	1/43	P	NOREV	117	F	??	WFE; With decal "LA JAMARI TANE".
☐ 180514-007	651 or 652	1/13	P	MONT BLANC	373125	F	76	WFE; Industrial tractor with front loader; Battery-electric motor; Steerable.
☐ 180514-008A	651-4	1/32	D	SOLIDO	510	F	78	WFE; Front-wheel-assist.
☐ 180514-008B	651-4	1/32	D	SOLIDO	510	F	79	WFE; Front-wheel-assist; Cab; Red, blue or orange; With or without four wheel wagon; Steering knob on top of cab.
☐ 180514-009	??	1/87	P	CURSOR	201	D	79	WFE; Orange.
☐ 180514-010	3000	1/32	D	MAJORETTE		F	82	WFE; With side mounted mower.
☐ 180514-011	145 14TX TURBO	1/43	D & P	YAXON	050	I	83	WFE; Front-wheel-assist; Cab; Decal variations on cab.
☐ 180514-012	145-14TX TURBO	1/32	D & P	YAXON	100	I	84	WFE; Front-wheel-assist; Cab.
☐ 180514-013	3040	1/30	CA	QUIRALU		F	56	WFE.
☐ 180514-014	3040	1/30	P	JOUEF		F	50	WFE.

	CODE	MODEL	SCALE	MATER.	MANU.	STOCK #	CTRY.	YEAR	REMARKS
☐	180514-015	86	1/12	P	CURSOR ?		D	70	WFE; Orange.
☐	180514-016	SUPER 5	1/87	D	CURSOR		D	66	WFE; Key chain (silver) made as a promotional for Renault Motoculture; Also without a keychain in orange.
☐	180514-017	??	1/12	P & T	CLIM		E	83	WFE; Yellow.
☐	180514-018	RK20CV	1/16	WOOD	CIJ		F	35	WFE; Rare.
☐	180514-019	MASTER	1/25	P	CURSOR		D	??	WFE.
☐	180514-020	1151/4	1/12	CUS.	LBS		F	??	WFE; Only a few made for Renault.
☐	180514-021A	651	1/13	P	MONT BLANC	323140	F	??	WFE; Cab; Promo.
☐	180514-021B	651	1/13	P	MONT BLANC	323140	F	??	WFE; Without cab; Remote control.
☐	180514-022	E-30	1/55	PK	CIJ		F	60	WFE.
☐	180514-023	TX 145-14	1/32	D	BRITAINS LTD.	9518	GB	85	WFE; Cab; Air cleaner steers tractor.

181519-000 ROSENGART

	CODE	MODEL	SCALE	MATER.	MANU.	STOCK #	CTRY.	YEAR	REMARKS
☐	181519-001	SUPER TRACTION	1/43	L	C.D.		F	32	WFE; Steam traction engine.

182113-000 RUMLEY

	CODE	MODEL	SCALE	MATER.	MANU.	STOCK #	CTRY.	YEAR	REMARKS
☐	182113-001A	OIL PULL (20-35)	1/16	CA	OLD TIME TOYS	NONE	USA	72	WFE; Represents a 1925 model; Serial numbered.
☐	182113-001B	OIL PULL (20-35)	1/16	CA	PIONEER TRACTOR WORKS	NONE	USA	81	WFE; Represents a 1925 model; PTW molded in; Serial numbered.
☐	182113-002	OIL PULL (20-30)	1/16	CA	ALVIN EBERSOL	NONE	USA	65	WFE; Represents a 1927 model.
☐	182113-003	OIL PULL	1/25	CA	IRVIN	NONE	USA	76	WFE; Represents an antique style tractor; Canopy, Driver.
☐	182113-004	OIL PULL (16-30)	1/16	CA	SCALE MODELS	#1	USA	80	WFE; Represents a 1919 ADVANCE RUMLEY OIL PULL; No. 1 in the JLE Thresher Series.
☐	182113-005	RUMLEY 6	1/10	WOOD	KRUSE	NONE	USA	83	WFE; "Steel" wheels.

190113-000 SAME

	CODE	MODEL	SCALE	MATER.	MANU.	STOCK #	CTRY.	YEAR	REMARKS
☐	190113-001	LEONE 70	1/65	P	POLISTIL	J-20	I	76	WFE; Mod-type tractor.
☐	190113-002	CENTAURO DT	1/15	D	DUGU	1	I	66	WFE; Front-wheel-assist; Excellent detail; Rare; DT-DOPPIA TRAZIONE.
☐	190113-003	LEONE 70	1/12	P	CO-MA	5131	I	71	WFE; Front-wheel-assist; Also a promo.
☐	190113-004A	BUFFALO 130	1/43	D & P	FORMA-PLAST	.085	I	77	WFE; Front-wheel-assist; Cab.
☐	190113-004B	BUFFALO 130	1/43	D & P	FORMA-PLAST		I	??	WFE; Front-wheel-assist; Cab; Sold by Mattel.
☐	190113-004C	BUFFALO 130	1/43	D & P	YAXON		I	??	WFE; Front-wheel-assist; Cab.
☐	190113-004D	BUFFALO 130	1/43	D	YAXON	0106	I	82	WFE; Front-wheel-assist; Cab; Loader.
☐	190113-005	PANTHER 50	1/20	P	VERVE	1016	I	80	WFE.
☐	190113-006	LEONE	1/65	P	POLISTIL	RJ 120	I	78	WFE.
☐	190113-007	TIGER 100	1/41	D & P	POLISTIL	CE 116	I	80	WFE; Front-wheel-assist; Cab.
☐	190113-008	TRIDENT 130	1/43	D	YAXON	084	I	83	WFE; Front-wheel-assist; Cab.
☐	190113-009	240 DT	1/16	D	??		I	61	WFE; Orange and brown; DT-DOPPIA TRAZIONE; Rare; Promo.
☐	190113-010	DA 38 DT	1/16	D	??		I	55	WFE; Front-wheel-assist; Orange and green; Clockwork; DT-DOPPIA TRAZIONE; Rare; Promo.
☐	190113-012	GALAXY 170	1/43	D	YAXON	085	I	84	WFE; Front-wheel-assist; Cab.
☐	190113-013	BUFFALO	1/30		??		NL	??	WFE; Promo.

190114-000 SATOH

	CODE	MODEL	SCALE	MATER.	MANU.	STOCK #	CTRY.	YEAR	REMARKS
☐	190114-001	BEAVER	1/32	P	??		J	78	WFE; Compact 15 horsepower tractor; Miniature has battery powered motor; With rear mounted roto-tiller.

190301-000 SCALE MODELS

	CODE	MODEL	SCALE	MATER.	MANU.	STOCK #	CTRY.	YEAR	REMARKS
☐	190301-001	SHOW TRACTOR	1/16	D	SCALE MODELS	NONE	USA	78	WFE; Many color and decal variations; Diecast or plastic seat variations.
☐	190301-002	SHOW TRACTOR	4"	P	SCALE MODELS	NONE	USA	82	Row crop; Many color and decal variations; Re-issue of Hubley tractor; Give-away at '82, '83, '84 toy shows with "SCALE MODELS MUSEUM 198" on decals.
☐	190301-003A	H	4"	D	SCALE MODELS	NONE	USA	82	Row crop; Reissue of Hubley H.
☐	190301-003B	H	4"	D	SCALE MODELS	NONE	USA	83	Same except Anniversary tractor for Joe and Helen Ertl's 25th Anniversary; Chrome plated.

190805-000 SHEPPARD DIESEL

	CODE	MODEL	SCALE	MATER.	MANU.	STOCK #	CTRY.	YEAR	REMARKS
☐	190805-001	SD-3	1/16	CA	SHEPPARD MFG. CO.	NONE	USA	50	Row crop; Represents a 1949 model; Only about 200 made.
☐	190805-002	SD-3	1/16	CA	A T & T COLLECTABLES	NONE	USA	83	Row crop; Copy of 190805-001.

190912-000 SILVER KING

	CODE	MODEL	SCALE	MATER.	MANU.	STOCK #	CTRY.	YEAR	REMARKS
☐	190912-001	??	1/16	CUS.	BEN SIEGEL	NONE	USA	83	Row crop; Single front wheel; Silver with red wheels.

190913-000 SIMPLICITY

	CODE	MODEL	SCALE	MATER.	MANU.	STOCK #	CTRY.	YEAR	REMARKS
☐	190913-001	GARDEN TRACTOR	1-1/4"	D	??		USA	??	Garden tractor; Gold color; Inside lucite prism.

191513-000 SOMECA

	CODE	MODEL	SCALE	MATER.	MANU.	STOCK #	CTRY.	YEAR	REMARKS
☐	191513-001	20	1/20	P	SEVITA		F	60	WFE.
☐	191513-002	35	1/20	P	SEVITA		F	60	WFE.
☐	191513-003	40	1/20	P	SEVITA		F	60	WFE.
☐	191513-004	640	1/36	D	DUGU		I	??	WFE; Three-point-hitch; Orange wheels; French version of Fiat.

	CODE	MODEL	SCALE	MATER.	MANU.	STOCK #	CTRY.	YEAR	REMARKS
☐	191513-005	640	1/36	D	OLD CARS	51	I	??	WFE; White wheels; French version of Fiat.
☐	191513-006	??	1/200	D	??		F	??	WFE; Enclosed in the cylinder of a keyring; Promo; Very rare.
☐	191513-007	??	1/55	P	BONUX		F	??	WFE.

192005-000 STEIGER

	CODE	MODEL	SCALE	MATER.	MANU.	STOCK #	CTRY.	YEAR	REMARKS
☐	192005-001	COUGAR II	1/12	CI	VALLEY PATTERNS		USA	75	WFE; Four-wheel-drive; Articulated; Cab; Cast without any engine; Only 100 made.
☐	192005-002A	BEARCAT III	1/12	CI	VALLEY PATTERNS		USA	76	WFE; Four-wheel-drive; Articulated; Cab; Cast without any engine; Only 150 made.
☐	192005-002B	PANTHER III	1/12	CI	VALLEY PATTERNS		USA	76	WFE; Four-wheel-drive; Articulated; Cab; Red, white and blue; Several decal variations; Cast without any engine.
☐	192005-003A	COUGAR III ST 251	1/32	P	ERTL	1930	USA	81	WFE; Four-wheel-drive; Articulated; Cab; Duals; "Classic Collectors Series" molded in; Red, white and blue.
☐	192005-003B	COUGAR III ST 251	1/32	D	ERTL	1930	USA	81	WFE; Four-wheel-drive; Articulated; Cab; Duals.
☐	192005-003C	PANTHER ST 310	1/32	P	ERTL	1925	USA	81	WFE; Four-wheel-drive; Articulated; Cab; Duals; "Classic Collectors Series" molded in.
☐	192005-003D	PANTHER ST 310	1/32	D	ERTL	1925	USA	81	WFE; Four-wheel-drive; Articulated; Cab; Duals.
☐	192005-003E	INDUSTRIAL	1/32	P	ERTL		USA	81	WFE; Four-wheel-drive; Articulated; Yellow; Duals; Yellow; "Classic Collectors Series" molded up.
☐	192005-004A	CP-1400 PANTHER 1000	1/16	CA	SCALE MODELS	1400	USA	83	WFE; Four-wheel-drive; Articulated; Cab; Duals; Serial numbered from 1 thru 1000.
☐	192005-004B	CP-1400 PANTHER 1000	1/16	CA	SCALE MODELS	1400	USA	83	WFE; Four-wheel-drive; Articulated; Cab; Duals.
☐	192005-005A	CP-1400 PANTHER 1000	1/32	D	SCALE MODELS	1900	USA	83	WFE; Four-wheel-drive; Articulated; Cab; Duals; Toy Farmer Special Issue Classic, 2000 made; Also made as a Steiger Special Issue; Chrome wheels.
☐	192005-005B	CP-1400 PANTHER 1000	1/32	D	SCALE MODELS	1900	USA	83	WFE; Four-wheel-drive; Articulated; Cab; Duals.
☐	192005-006	PANTHER 1000 *	1/12	CI	VALLEY PATTERNS		USA	83	WFE; Four-wheel-drive; Articulated; Cab; Duals; The following decal variations exist: CP 1325, CP 1360, CP 1400, KP 1325, KP 1360 and KP 1400.
☐	192005-007A	COUGAR	1/64	D	ERTL	1945	USA	82	WFE; Four-wheel-drive; Articulated; Cab.
☐	192005-007B	INDUSTRIAL	1/64	D	ERTL	1980	USA	84	WFE; Four-wheel-drive; Articulated; Cab; Industrial yellow.
☐	192005-008A	CP-1400 PANTHER 1000	1/64	D	SCALE MODELS	1910	USA	84	WFE; Four-wheel-drive; Articulated; Cab; Duals; "First Edition" molded in.
☐	192005-008B	CP-1400 PANTHER 1000	1/64	D	SCALE MODELS	1910	USA	85	WFE; Four-wheel-drive; Articulated; Cab; Duals.
☐	192005-009	PANTHER CP-1400	1. Liter	P	PACESETTER	No. 3	USA	84	WFE; Four-wheel-drive; Articulated; Cab; Duals; Decanter; "Big Panther".
☐	192005-010	PANTHER CP-1400	200 ML	POR	PACESETTER	No. 3	USA	84	WFE; Four-wheel-drive; Articulated; Cab; Duals; Decanter; "Big Panther".
☐	192005-011A	PANTHER IV FIRST EDITION	1/32	D	SCALE MODELS	D	USA	84	WFE; Four-wheel-drive; Articulated; Cab.
☐	192005-011B	PANTHER IV	1/32	D	SCALE MODELS		USA	84	WFE; Four-wheel-drive; Articulated; Cab.
☐	192005-012	PANTHER SERIES II	1/16	CUS.	DAVID SHARP	NONE	USA	80	WFE; Four-wheel-drive; Cab; Duals; Articulated.
☐	192005-013	SERIES II	1/16	CUS.	GARY ANDERSON	NONE	USA	84	Four-wheel-drive; Cab; Duals; Articulated.

192018-000 STEYR

	CODE	MODEL	SCALE	MATER.	MANU.	STOCK #	CTRY.	YEAR	REMARKS
☐	192018-001A	8160a	1/43	D & P	YAXON	059	I	79	WFE; Four-wheel-drive; Cab; Red and white.
☐	192018-001B	8160b	1/43	D & P	YAXON	0118	I	79	WFE; Four-wheel-drive; Cab; Red and white; Duals.
☐	192018-001C	8170	1/43	D & P	YAXON	059	I	84	WFE; Four-wheel-drive; Cab.
☐	192018-002	180a, 80 or 280	1/30	P	KUNSTOFF HERBERT FITZEK		A	56	WFE; Gray or green; Name "Steyr" on wheel centers.
☐	192018-003	??	1/32	T	G.I.H.	290	D	??	Crawler with trailer; Has the Steyr logo.

192301-000 SWARAJ

	CODE	MODEL	SCALE	MATER.	MANU.	STOCK #	CTRY.	YEAR	REMARKS
☐	192301-001A	(735)	1/20	D	MAXWELL		IN	82	WFE; Green and white.
☐	192301-001B	735	1/20	D	MAXWELL		IN	83	WFE; Green and white; Has number "735" on decals.

200112-000 TALYOR

	CODE	MODEL	SCALE	MATER.	MANU.	STOCK #	CTRY.	YEAR	REMARKS
☐	200112-001	CLIPPER 1882	1/43	CUS.	PHANTOM MODELS	A06	F	??	Portable gasoline engine. Represents an antique engine.

200518-000 TERRATRAK

	CODE	MODEL	SCALE	MATER.	MANU.	STOCK #	CTRY.	YEAR	REMARKS
☐	200518-001	??	6-1/2"	CA	PLOW BOY TOYS	NONE	USA	79	Crawler; Yellow.

CODE	MODEL	SCALE	MATER.	MANU.	STOCK #	CTRY.	YEAR	REMARKS
200913-000	**TIMBERJACK**							
☐ 200913-001	240 E	1/12	P	PROMOTION G. BELANGER	NONE	CDN	83	WFE; Four-wheel-drive; Log skidder; Articulated; Blade; ROPS; Orange.

NOW A PART OF INTERNATIONAL HARVESTER CORP.

CODE	MODEL	SCALE	MATER.	MANU.	STOCK #	CTRY.	YEAR	REMARKS
200920-000	**TITAN**							
☐ 200920-001	OIL TRACTOR	1/16	CA	SCALE MODELS	NO. 4	USA	82	WFE; Represents a 1912 model; No. 4 in the JLE Threshers Series.
201725-000	**TOYOTA**							
☐ 201725-001	??	1/50	D	TOMICA	96	J	??	WFE; Airport tractor.
☐ 201725-002	JOBSUN SDK8	1/54	D	TOMICA	37	J	??	Four-wheel-skid steer loader.

MARSHALL

CODE	MODEL	SCALE	MATER.	MANU.	STOCK #	CTRY.	YEAR	REMARKS
201801-000	**TRACK MARSHALL**							
☐ 201801-001	135	1/32	WM KIT	SCALEDOWN		GB	83	Crawler; Cab; Excellent detail including individual metal track sections.
202107-000	**TUGSTER**							
☐ 202107-001	TYPE 37A	6"	D	TRIANG		GB	??	WFE; Green or blue.

NOW A PART OF WHITE FARM EQUIPMENT VIA MINNEAPOLIS-MOLINE.

CODE	MODEL	SCALE	MATER.	MANU.	STOCK #	CTRY.	YEAR	REMARKS
202309-000	**TWIN CITY**							
☐ 202309-001	(60-90)	1/16	K	ROBERT GRAY	NONE	USA	74	WFE; "Steel wheels"; Represents a 1918 model; Largest internal combustion engine tractor ever made.
211819-000	**URSUS**							
☐ 211819-001	??	1/43	P	ZTS PLASTIK		POL	81	WFE; Red; Sold with small trailer.
☐ 211819-002	C 335	1/20	P	??		POL	71	WFE; Steerable; Red and gray with cream colored wheels; Three-point-hitch; Very good detail.
☐ 211819-003	C 385	1/43	??	ESTETYCA		POL	80	WFE.
☐ 211819-004	??	1/10		??		POL		WFE; Old style tractor.
211820-000	**URTRAK**							
☐ 211820-001A	KS 30	1/87	P	ESPEWE	1017/5	DDR	??	Agricultural crawler.
☐ 211820-001B	KS 30	1/87	P	ESPEWE	1017/1	DDR	??	Military crawler.
☐ 211820-001C	KT 50	1/87	P	ESPEWE	1017/3	DDR	??	Bulldozer.
☐ 211820-001D	KT 50	1/87	P	ESPEWE	1017/4	DDR	??	Military bulldozer.
☐ 211820-001E	KT 50 UK	1/87	P	ESPEWE	1017/2	DDR	??	Crawler-loader.
220112-000	**VALMET**							
☐ 220112-001	138-4	1/50	D	ARPRA	35	BR	84	WFE; Front-wheel-assist; ROPS/canopy; Three-point-hitch.
220514-000	**VENDEUVRE**							
☐ 220514-001	BL	1/16	D	VENDEUVRE		F	60	WFE; The Vendeuvre Company was acquired by Allis Chalmers in 1962.
220518-000	**VERSATILE**							
☐ 220518-001A	895	1/16	CA	SCALE MODELS		USA	80	WFE; Four-wheel-drive; Cab; Articulated; Duals; Yellow and red; Metal or rubber air cleaner and exhaust stacks.
☐ 220518-001B	895	1/16	CA	SCALE MODELS		USA	82	WFE; Same except having external planatary gears on the wheels.
☐ 220518-001C	895	1/16	CA	SCALE MODELS		USA	83	WFE; Four-wheel-drive; Articulated; Cab; Color change and "Designation 6" decals (six different model numbers on a strip to customize model.
☐ 220518-002A	1150	1/16	CA	SCALE MODELS		USA	82	WFE; Four-wheel-drive; Cab; Articulated; Triple wheels on front and rear; 400 mounted on walnut plaques for Versatile dealer presentations.
☐ 220518-002B	1150	1/16	CA	SCALE MODELS	9900328	USA	83	WFE; Same except darker yellow color and different grill and trim.
☐ 220518-002C	1150	1/16	CA	SCALE MODELS		USA	82	WFE; Four-wheel-drive; Cab; Articulated; Mounted on walnut plaque for Versatile dealers convention, November 1982.
☐ 220518-003A	1150	1/32	D	SCALE MODELS	99000291	USA	83	WFE; Four-wheel-drive; Cab; Articulated; Triples front and rear; No stops to prevent too much articulation (Play safety hazard).
☐ 220518-003B	1150	1/32	D	SCALE MODELS	9900029	USA	83	WFE; Four-wheel-drive; Cab; Articulated; Triples front and rear; Has articulation stops; Darker yellow.
☐ 220518-004A	256 FIRST EDITION	1/32	D	SCALE MODELS	99000340	USA	84	WFE; Four-wheel-drive; Cab; Articulated; Collectors Edition.
☐ 220518-004B	256	1/32	D	SCALE MODELS	99000340	USA	84	WFE; Four-wheel-drive; Cab; Articulated.
☐ 220518-004C	276	1/32	D	SCALE MODELS		USA	84	WFE; Four-wheel-drive; Cab; Articulated; With attachments.
☐ 220518-005	825	1/16	CA	SCALE MODELS		USA	84	WFE; Four-wheel-drive; Cab; Articulated; Duals; Represents an older model.
☐ 220518-006A	836	1/32	D	SCALE MODELS		USA	84	WFE; Four-wheel-drive; Cab; Articulated; Commemorates the "World Premier Showing".
☐ 220518-006B	836	1/32	D	SCALE MODELS		USA	84	WFE; Four-wheel-drive; Cab; Articulated; "First Edition".
☐ 220518-006C	836	1/32	D	SCALE MODELS		USA	84	WFE; Four-wheel-drive; Cab; Articulated; Regular issue.
☐ 220518-007	935	1/16	CUS.	DAVE SHARP	NONE	USA	80	WFE; Four-wheel-drive; Articulated; Cab; Duals.

	CODE	MODEL	SCALE	MATER.	MANU.	STOCK #	CTRY.	YEAR	REMARKS
☐	220518-008	BIG ROY	1/16	CUS.	DAVE SHARP	NONE	USA	82	WFE; Eight-wheel-drive; Four axles; Articulated; Experimental model at the Versatile Company.

220903-000 VICKERS

	CODE	MODEL	SCALE	MATER.	MANU.	STOCK #	CTRY.	YEAR	REMARKS
☐	220903-001	VIGOR	1/16	P	VIPRODUCT VICTORY IND.		GB	52	Crawler; Battery electric; Very good detail; Yellow or blue; Rare.

220905-000 VIERZON

	CODE	MODEL	SCALE	MATER.	MANU.	STOCK #	CTRY.	YEAR	REMARKS
☐	220905-001	401	1/30	D	JRD		F	50	WFE; Rare.

230112-000 WALLIS

	CODE	MODEL	SCALE	MATER.	MANU.	STOCK #	CTRY.	YEAR	REMARKS
☐	230112-001	(20-30)	1/25	CI	VINDEX		USA	27	WFE; Rare.
☐	230112-002	(20-30)	1/16	CA	SCALE MODELS	No. 6	USA	81	WFE; "Steel" wheels; Represents a 1933 model; No. 6 in the JLE Collector Series; 3000 made.

230501-000 WEATHERHILL

	CODE	MODEL	SCALE	MATER.	MANU.	STOCK #	CTRY.	YEAR	REMARKS
☐	230501-001	(12H)	1/57	D	LESNEY	KI	GB	??	WFE; Industrial tractor with front loader.
☐	230501-002	??	1/88	D	LESNEY	24	GB	??	WFE; Industrial tractor with front loader.
☐	230501-003	??	1/75	D	LESNEY	24	GB	??	WFE; Industrial tractor with front loader.
☐	230501-004	??	1/75	D	JADALI		F	??	WFE; Industrial tractor with front loader; Copy of Lesney.

230519-000 WESTRAK

	CODE	MODEL	SCALE	MATER.	MANU.	STOCK #	CTRY.	YEAR	REMARKS
☐	230519-001	??	7"	CA	PLOW BOY TOYS	NONE	USA	79	Crawler; Yellow.

230809-000 WHITE or WFE — CONSOLIDATED FROM OLIVER, MINNEAPOLIS-MOLINE AND COCKSHUTT.

	CODE	MODEL	SCALE	MATER.	MANU.	STOCK #	CTRY.	YEAR	REMARKS
☐	230809-001A	2-135	1/16	D	SCALE MODELS		USA	78	WFE; Cab; Silver decals; Some were mounted on walnut plaques a commemorative models; Rubber muffler; Formerly listed as Dyersville Die-Cast.
☐	230809-001B	2-155	1/16	D	SCALE MODELS		USA	78	WFE; Cab; Dual wheels; Rubber muffler; Formerly listed as Dyersville Die-Cast.
☐	230809-001C	2-135 FWA	1/16	D	SCALE MODELS		USA	79	WFE; Front-wheel-assist; Cab; Rubber muffler.
☐	230809-001D	2-135	1/16	D	SCALE MODELS		USA	80	WFE; Same except no muffler.
☐	230809-001E	2-135 FWA	1/16	D	SCALE MODELS		USA	80	WFE; Front-wheel-assist; Same except no muffler.
☐	230809-001F	2-155	1/16	D	SCALE MODELS		USA	80	WFE; Duals; Same except no muffler.
☐	230809-001G	2-135	1/16	D	SCALE MODELS		USA	81	WFE; Same except with muffler-aspirator and red decals.
☐	230809-001H	2-135 FWA	1/16	D	SCALE MODELS		USA	81	WFE; Front-wheel-assist; Same except with muffler-aspirator and red decals.
☐	230809-001I	2-155	1/16	D	SCALE MODELS		USA	81	WFE; Duals; Same except with muffler-aspirator and red decals.
☐	230809-002A	2-180	1/16	D	SCALE MODELS		USA	79	WFE; Cab; Dual rear wheels; Caterpillar engine; No muffler; Silver decals.
☐	230809-002B	2-180	1/16	D	SCALE MODELS		USA	82	WFE; Same except with muffler-aspirator and red decals.
☐	230809-003A	2-35 ISEKI	1/25	D	SCALE MODELS	23579	USA	79	WFE; Silver decals; Fenders without rivet.
☐	230809-003B	2-35 ISEKI	1/25	D	SCALE MODELS	23579	USA	83	WFE; Red decals; Fenders riveted on.
☐	230809-004	4-210	1/16	D	SCALE MODELS	421080	USA	80	WFE; Four-wheel-drive; Articulated; Cab; Silver decals.
☐	230809-005A	4-175	1/16	D	SCALE MODELS		USA	81	WFE; Four-wheel-drive; Articulated; Cab; Silver decals; Very limited production.
☐	230809-005B	4-175	1/16	D	SCALE MODELS		USA	81	WFE; Same except red decals.
☐	230809-006	(JUNIOR)	1/25	D	SCALE MODELS		USA	83	Row crop; Non-steering; Used as a commemorative tractor for such events as The Louisville, KY 2/15-18/84 Show.
☐	230809-007	4-225	1/16	P	SCALE MODELS		USA	83	WFE; Four-wheel-drive; Articulated; Cab.
☐	230809-008A	4-270 Comm.	1/16	CA	SCALE MODELS	4-270	USA	84	WFE; Four-wheel-drive; Articulated; Cab; Duals; New fender fuel tanks; Introduced at The National Farm Machinery Show, Louisville, KY.
☐	230809-008B	4-270	1/16	CA	SCALE MODELS	4-270	USA	84	WFE; Same except without commemorative inscription.
☐	230809-009A	2-135 FIRST EDITION	1/64	P	SCALE MODELS		USA	84	WFE; Cab; Collectors version.
☐	230809-009B	2-135	1/64	P	SCALE MODELS		USA	85	WFE; Cab.
☐	230809-009C	2-155	1/64	P	SCALE MODELS	NONE	USA	85	WFE; Duals; Cab.
☐	230809-009D	2-180	1/64	P	SCALE MODELS	NONE	USA	85	WFE; Duals; Cab.
☐	230809-010	700	1/16	D & P	SCALE MODELS	700	USA	84	WFE; "7TH ANNUAL DYERSVILLE SHOW" tractor.
☐	230809-011	(2-135)		CER	??		USA	82	WFE; Ceramic flower pot.
☐	230809-012	4-180	1/16	CUS.	DAVID SHARP	NONE	USA	81	WFE; Four-wheel-drive; Cab; Articulated.

230812-000 WHEEL HORSE

	CODE	MODEL	SCALE	MATER.	MANU.	STOCK #	CTRY.	YEAR	REMARKS
☐	230812-001	LGT	1/24	PK	MPC		USA	??	Lawn and garden tractor; Came with an Indy racing car as a set.

CODE	MODEL	SCALE	MATER	MANU.	STOCK #	CTRY.YEAR	REMARKS
230820-001 WHITLOCK							
☐ 230820-001	??	1/30	P	??	No. 78	DK 50	Crawler; Red.
231512-000 WOLSELEY							
☐ 231512-001	WD2	1/15	WM KIT	SCALEDOWN	E3A	GB ??	Small portable gasoline engine—1-1/2 HP.
250112-000 YALE-EATON						**See also Eaton-Yale.**	
☐ 250112-001	040R	1/25	D	CONRAD	299	D 80	Industrial fork lift.
250114-000 YANMAR							
☐ 250114-001	TB-20	7"	P	T-N	230	J ??	WFE; Friction motor; Incorrectly listed previously as a Porsche.
☐ 250114-002	YM 3110	1/25	D	DIAPET	127-01481 T-78	J 79	WFE; Tipping hood; With rear mounted roto-tiller.
260520-000 ZETOR HMT							
☐ 260520-001	5511	1/80	D	CORGI (JR)	4	GB70	WFE; Cab; Orange or green.
☐ 260520-002	SUPER DIESEL	1/40	P	U.S.U.D.	KCS-11	CS 73	WFE; Friction motor; Rare.
☐ 260520-003	HMT 2511	1/25	D	MAXWELL	570	IN 80	WFE; Red.
☐ 260520-004	??	1/87	D	CHARMERZ		HK 79	WFE; With livestock trailer.
☐ 260520-005A	CRYSTAL 12045	1/43	P	IGRA		CZ 81	WFE; Front-wheel-assist; Cab.
☐ 260520-005B	CRYSTAL 12045	1/43	PK	IGRA		CZ 83	WFE; Front-wheel-assist; Cab.
☐ 260520-005C	CRYSTAL 12045	1/43	P	IGRA		CZ 84	WFE; Front-wheel-assist; Cab; With front loader.
☐ 260520-006	(5511)	1/80	D	PLAYART	7170	HK 75	WFE; Cab; Green, Copy of Corgi Jr. (260520-001).
☐ 260520-007	SUPER	1/25	P	MACHPI CASCHI		I 57	WFE; Red.
☐ 260520-008A	25	1/20	T	KDN		CS 65	WFE; Clockwork; Sectional wheel weights; Open grill.
☐ 260520-008B	25	1/20	T	KDN		CS 67	WFE; Clockwork; Round wheel weights; Closed grill.
☐ 260520-009	CRYSTAL 8011	1/18	P & T	KDN		CS 83	WFE; Remote control; Red or blue.
260525-000 ZETTLEMEYER							
☐ 260525-001	DOZER ZD 3000 IBH	1/50	D	NZG	195	D 81	Industrial crawler.
☐ 260525-002	LADER EUROP L 2000	1/50	D	SIKU	V270	D 67	Industrial loader; Siku Super Series Zinkguss.

Pedal Tractor Charts

CODE	MODEL	SCALE	MATER.	MANU.	STOCK #	CTRY.	YEAR	REMARKS
PT0103-000	**ALLIS CHALMERS**							
☐ PT0103-001	(C)	34-1/2"	CA	ESKA		USA	49-	
☐ PT0103-002	CA	38-1/2"	CA	ESKA		USA	50	Has "CA" designation.
☐ PT0103-003	D-14	38"	CA	ESKA		USA	57	Cast-in grill design.
☐ PT0103-004	D-17	39"	CA	ESKA		USA	58	Smooth grill with decal.
☐ PT0103-005	190	37"	CA	ERTL	AC-190	USA	64	Bar grill; Casting code A-64.
☐ PT0103-006A	190 XT	37"	CA	ERTL	AC-190	USA	66	No bars on grill; 190 winged decal on hood; Casting code A-64.
☐ PT0103-006B	190 XT	37"	CA	ERTL		USA	67	Same casting, later decal design with "XT" designation.
☐ PT0103-007	200	37"	CA	ERTL		USA	72	Casting code A-64.
☐ PT0103-008	7080	37"	CA	ERTL		USA	75	Maroon and orange; Casting code A-64.
☐ PT0103-009	7045	37"	CA	ERTL		USA	78	Orange and black; Casting code A-64.
☐ PT0103-010	8070	37"	CA	ERTL		USA	82	Orange and black; Casting code A-64.
PT0301-000	**CASE**							
☐ PT0301-001	(VAC)	35"	CA	ESKA		USA	53-	
☐ PT0301-002	(400)	38-1/2"	CA	ESKA		USA	55-	
☐ PT0301-003	CASE-O-MATIC	38"	CA	ESKA		USA	58-	
☐ PT0301-004	PLEASURE KING 30	39"	CA	ERTL		USA	65-	
☐ PT0301-005	1070 AGRI-KING	35-1/2"	CA	ERTL		USA	70	Desert sunset (sand) and power red; Casting code C-63.
☐ PT0301-006	CASE AGRI-KING	35-1/2"	CA	ERTL		USA	73	Power red and white; Casting code C-63.
☐ PT0301-007	CASE	35-1/2"	CA	ERTL		USA	80	Power red and white; Casting code C-63.
PT0318-000	**CATERPILLAR**							
☐ PT0318-001A	D-4	36" without	SS	NEW LONDON METAL PRODUCTS		USA	49	With dozer blade; Pedal variation.
☐ PT0318-001B	D-4	36" without blade	SS	NEW LONDON METAL PRODUCTS		USA	49	With dozer blade; Battery powered variation.
☐ PT0318-001C	D-4	36" without blade	SS	NEW LONDON METAL PRODUCTS		USA	49	With dozer blade; Gasoline engine powered.
PT0402-000	**DAVID BROWN**							
☐ PT0402-001	DAVID BROWN		Plastic	SHARNA		GB	69	Power red, orchid white and chocolate colors.
PT0405-000	**DEUTZ**							
☐ PT0405-001A	5006	35"	Plastic	ROLLY TOYS		D	??	Dark green and gray.
☐ PT0405-001B	5006	35"	Plastic	ROLLY TOYS		D	??	Light green and gray.
☐ PT0405-002	4.70	31"	Plastic	ROLLY TOYS		F	??	First model to designated Deutz-Fahr merger.
☐ PT0405-003	DX 110	80cm	Plastic	ROLLY TOYS	490	D	79-	
☐ PT0405-004A	DX 110	100cm 38-1/2"	Plastic	ROLLY TOYS	290	D	79-	
☐ PT0405-004B	DX 110	38-1/2" without loader	Plastic	ROLLY TOYS	289	D	79	Same except with front end loader; Overall length 128cm.
☐ PT0405-005	DX 85	22-1/2"	Plastic	FALK		F	81	Tricycle front end.
☐ PT0405-006	DX 160	74cm 29-1/2"	Plastic	FALK	762	F	79	Cleated front wheels.
☐ PT0405-007	230	90cm	Plastic	FALK		F	??-	
PT0605-000	**FENDT**							
☐ PT0605-001	FAVORIT TURBOMATIK	39-1/2"	Plastic	ROLLY TOYS		D	??-	
PT0609-000	**FIAT**							
☐ PT0609-001	880		Plastic	FUCHS	5529	D	??	Orange with Fiat markings.
☐ PT0609-002	FIAT	80cm	Plastic			F	77-	
PT0615-000	**FORD**							
☐ PT0615-001	(900)	40"	CA	GRAPHIC REPRODUCTION		USA	54	Gray and red; Either tin or cast aluminum fenders; Recessed vertical grooved grill.
☐ PT0615-002	(901)	40"	CA	GRAPHIC REPRODUCTION		USA	58	Gray and red; Either tin or cast aluminum fenders; Grid-type grill bars.
☐ PT0615-003	COMMANDER 6000	38-1/2"	CA	ERTL		USA	65	Blue and gray; Casting code F-65.
☐ PT0615-004	8000	36"	CA	ERTL		USA	??	Casting code F-68.
☐ PT0615-005	TW-20	36"	CA	ERTL		USA	80	Casting code F-68.
☐ PT0615-006	7700		Plastic	FUCHS		D	??	
☐ PT0615-007	(4000)	130cm	Plastic	??		F	65	Two piece grill style.
☐ PT0615-008	TW-20	75cm (35")	Plastic	FALK		F	80	
☐ PT0615-009	FORD	110cm 24-1/2"	Plastic	FALK		F	??	Tricycle.
☐ PT0615-010	TW-10	27" with tipper dump	Plastic	FALK		F	??	Has rear tipping box.
☐ PT0615-011	TW-30	90cm	Plastic	FALK		F	??	
☐ PT0615-012	(4000)		Plastic	TRI-ANG		GR	75	

PT0908-000 INTERNATIONAL

CODE/MODEL	SCALE	MATER.	MANU.	STOCK #	CTRY.	YEAR	REMARKS
PT0908-001 (H)	33"	CA	ESKA		USA	49	Open grill; Cast aluminum seat; Small 1" x 6" front wheels; Smooth rear tires; Larger cast-in muffler; Motor detail varies from -002.
PT0908-002 (H)	35-1/2"	CA	ESKA		USA	49	Closed grill; Steering rod support is a separate casting.
PT0908-003 (H)	33"	CA	ESKA		USA	50	High steering post; Closed grill.
PT0908-004 (M)	35-1/2"	CA	ESKA		USA	50	Noise maker; Pressed steel seat; No muffler or air cleaner stack.
PT0908-005 400	39"	CA	ESKA		USA	54	Cast-in grill design.
PT0908-006 450	39"	CA	ESKA		USA	56	Grill decal.
PT0908-007 560	38"	CA	ESKA		USA	58	
PT0908-008 806	36-1/2"	C	ERTL		USA	63	Casting I-64.
PT0908-009 856	36-1/2"	CA	ERTL		USA	67	Casting code I-64.
PT0908-010 1026 HYDRO	36-1/2"	CA	ERTL		USA	70	Casting number I-64.
PT0908-011 66 SERIES	37"	CA	ERTL		USA	71	Casting number 404.
PT0908-012 86 SERIES	37"	CA	ERTL		USA	80	Casting number 404.
PT0908-013 INTERNATIONAL 844		Plastic	ROLLY TOYS		D	78	Wide front axle; European utility style.
PT0908-014 INTERNATIONAL 844	65cm	Plastic	ROLLY TOYS	561	D	??	Row crop front end.
PT0908-015 1488 XL		Plastic	FALK		F	??	Wide front end.
PT0908-016 INTERNATIONAL 1055	91cm	Plastic	FALK		F	79	
PT0908-017 633	24-1/2"	Plastic	ROLLY TOYS		D	??	Tricycle front end.
PT0908-018 1455XL	36"	Plastic	FALK		F	??	
PT0908-019 955	75cm	Plastic	FALK		F	??	

PT1004-000 JOHN DEERE

CODE/MODEL	SCALE	MATER.	MANU.	STOCK #	CTRY.	YEAR	REMARKS
PT1004-001 (A)	33-1/2"	CA	ESKA		USA	50	Open grill.
PT1004-002 (60)	34"	CA	ESKA		USA	52	Steering wheel and rear hitch area casting variations; This version had round spherical hub caps.
PT1004-003 (60)	38"	CA	ESKA		USA	55	No exhaust stack; Flat hub caps.
PT1004-004 (620)	37-1/2"	CA	ESKA		USA	56	
PT1004-005A(130)	38"	CA	ESKA		USA	58	One bolt holding front assembly together; Open casting area above fan drive shaft.
PT1004-005B(130)	38"	CA	ESKA		USA	58	Casting variation open above fan drive shaft; Used two bolts to fasten casting together above front wheels.
PT1004-006A10	37"	CA	ESKA		USA	61	Three holes on left side of engine casting; Represents 4010 Series; Eska tires.
PT1004-006B10	37"	CA	ESKA		USA	63	Four holes on left side of engine casting; Represents 4010 Series.
PT1004-007A20	38"	CA	ERTL		USA	63	Represents 4029 Series; Smaller rear housing for sprocket; Casting code D-63.
PT1004-007B20	38"	CA	ERTL		USA	65	Similar to 007A except rear housing is larger; Casting code D-6
PT1004-008 LGT	35-1/2"	CA	ERTL		USA	??	Front hub cover variations, either large or small hub caps; Casting code DTG-70.
PT1004-009 (4430)	37"	CA	ERTL		USA	73	Casting number 520.
PT1004-010 (4440)	37"	CA	ERTL		USA	78	Casting code 520.
PT1004-011 (4450)	37"	CA	ERTL		USA	82	Casting number 520.
PT1004-012A3140	40"	Plastic	ROLLY TOYS	295	D	??	
PT1004-012B3140	40" without loader	Plastic	ROLLY TOYS	292	D	??	Same but with front loader.

PT1306-000 MASSEY-FERGUSON

CODE/MODEL	SCALE	MATER.	MANU.	STOCK #	CTRY.	YEAR	REMARKS
PT1306-001 (1130)	36-1/3"	CA	ERTL		USA	76	Casting number 1100.
PT1306-002 (2775)	36"	CA	ERTL		USA	76	Casting code 1100.
PT1306-003 188	77cm 30-1/2"	Plastic	ROLLY TOYS	291	D	??	Smaller version of PT1306-004A; No fenders.
PT1306-004 (188)	96cm 37-1/2"	Plastic	ROLLY TOYS	286	D	??	Without front loader.
PT1306-005 145	92cm (36")	Plastic	FALK		F	82	"Multi-Power" decal; Blue fenders.
PT1306-100		Plastic	FALK		F	77	Tricycle front end.
PT1306-007 2640	30" M	Plastic	FALK		F	83	Plastic with cleted front wheels.

PT1308-000 MASSEY-HARRIS

CODE/MODEL	SCALE	MATER.	MANU.	STOCK #	CTRY.	YEAR	REMARKS
PT1308-001 (44)	33"	CA	??	DA-116	USA	47	Open grill; Features "Velvet Ride" spring mounted seat; Air cleaner and muffler stacks on hood.
PT1308-002 44 SPECIAL	38-1/2"	CA	ESKA		USA	53	

PT1309-000 MISCELLANEOUS

CODE/MODEL	SCALE	MATER.	MANU.	STOCK #	CTRY.	YEAR	REMARKS
PT1309-001 SUPER S 150		Plastic	FUCHS	5524	D	??	Red plastic body.
PT1309-002 ??	96cm	Plastic	FERBEDO		D	81	Tractor with blinking lights; Light green and silver.
PT1309-003 285	92cm	Plastic	ROLLY TOYS	285	D	??	Blue and white; Plastic; A non-descript model.

CODE	MODEL	SCALE	MATER.	MANU.	STOCK #	CTRY.	YEAR	REMARKS
☐	PT1309-004 ROLLY TRAKTOR	124cm	Plastic	ROLLY TOYS	484	D	??	With front end loader; Blue and red.
☐	PT1309-005 BIG 4	39-1/2"	SS	AMT		USA	??	Pressed steel construction; Chain drive; Wide front axle.
☐	PT1309-006 HUFFY					USA	??	
☐	PT1309-007 FARMER BOY			??		USA	??	Belt driven; Cast aluminum hood and grill with raised letters.
☐	PT1309-008 BIG		Plastic	??		D	??	Green plastic with red seat and wheels; White steering wheel.
☐	PT1309-009A BMC HEAVY TRACTOR JUNIOR	36"	SS	BINGHAMTON MFG. CORP.		USA	48	Push pedals; Made in Binghamton, New York; Implements also available.
☐	PT1309-009B BMC HEAVY TRACTOR SENIOR	40"	SS	BINGHAMTON MFG. CORP.		USA	48	Belt drive; Made in Binghamton, New York; Implements also available.
☐	PT1309-010A TRACTALL CUB	41-1/2"	CA & S	INLAND MFG. CO.		USA	??	Early version has cast aluminum hood with raised letters "Tractall" on hood; Belt drive; "Cub" size was for smaller children.
☐	PT1309-010B TRACTALL CUB	41.4"	SS	INLAND MFG. CO.		USA	??	Later model had stamped steel hood with printed lettering including letter "C"; Chain drive.
☐	PT1309-011A TRACTALL HUSKY	LARGE	CA & S	INLAND MFG. CO.		USA	??	Early version had cast aluminum hood with raised letters on hood "Tractall"; Belt drive; "Husky" model was for larger children.
☐	PT1309-011B TRACTALL HUSKY	LARGE	SS	INLAND MFG. CO.		USA	??	Later model had stamped steel hood with printed lettering (C); Chain drive; Husky size for larger children.
☐	PT1309-012 ROADMASTER	39-1/2"	PL & METAL	ROADMASTER CORP.		USA	84	Same as Wheel Horse (PT2308-001) but with Roadmaster decals.

PT1313-000   MINNEAPOLIS-MOLINE

CODE	MODEL	SCALE	MATER.	MANU.	STOCK #	CTRY.	YEAR	REMARKS
☐	PT1313-001 TOT TRACTOR	40"	SS	BMC		USA	??	Belt drive; Large red MM decal on each side; "Tot Tractor" on base of decal.
☐	PT1313-002 ??	41"	SS	BMC		USA	??	Chain drive; Fenders; Bullet shaped front grill; Shuttle shift.

PT1512-000   OLIVER

CODE	MODEL	SCALE	MATER.	MANU.	STOCK #	CTRY.	YEAR	REMARKS
☐	PT1512-001 88	33"	CA	ESKA		USA	47	Open grill.
☐	PT1512-002 88	33"	CA	ESKA		USA	47	Closed grill.
☐	PT1512-003 SUPER 88	LARGE	CA	ESKA		USA	54	
☐	PT1512-004 880 DIESEL	39"	CA	ESKA		USA	58	Grill decal.
☐	PT1512-005 1800 A	39"	CA	ESKA		USA	62	Decal with keystone on sides; Solid cast-in grill with decal.
☐	PT1512-006 1800 C	39-1/2"	CA	ERTL	1004	USA	63	Red border side decals; Plastic grill; Casting code O-63.
☐	PT1512-007 1850	39-1/2"	CA	ERTL		USA	64	Casting code O-63.
☐	PT1512-008 1855	39-1/2"	CA	ERTL		USA	69	Casting code O-63.
☐	PT1512-009 1855 WHITE	39-1/2"	CA	ERTL		USA	72	Casting code O-63; This is the last one in the Oliver pedal tractor series. The "White" name on the decal indicated the merger forming the White Corporation.

PT1805-000   RENAULT

CODE	MODEL	SCALE	MATER.	MANU.	STOCK #	CTRY.	YEAR	REMARKS
☐	PT1805-001 1451-4	83cm (33")	Plastic	AMPA FRANCE	63162A	F	81	Vertical exhaust; Cleted front wheels.
☐	PT1805-002 TURBO TX 1.33-14	75cm 29-1/2"	Plastic	FALK		F	??	
☐	PT1805-003 RENAULT	110cm	Plastic	FALK		F	??	Tricycle front end.
☐	PT1805-004 RENAULT TURBO	86cm	Plastic	AMPA FRANCE		F	??	Battery powered; Yellow.
☐	PT1805-005 951-4	82cm (36")	Plastic	FALK		F	76	Vertical exhaust; Cleted front wheels; "Tracto-Control" featured on hood decal.
☐	PT1805-006 751-4	29"	Plastic	FALK		F	??	

PT2215-000   VOLVO

CODE	MODEL	SCALE	MATER.	MANU.	STOCK #	CTRY.	YEAR	REMARKS
☐	PT2215-001 VOLVO BM		Plastic	FUCHS	5536	D	??	

PT2308-000   WHEEL HORSE

CODE	MODEL	SCALE	MATER.	MANU.	STOCK #	CTRY.	YEAR	REMARKS
☐	PT2308-001 C-195	39-1/2"	PL & METAL	ROADMASTER CORP.	10A21 G439	USA	84	Chain drive; Adjustable seat.

Glossary

Air Cleaner Stack. An extension of the air cleaner up through or along side the hood.

Antique. Refers to very old style tractors made in the 1930's or earlier.

Articulated

Articulated. A machine that is designed to bend in the middle for the purpose of steering. Generally these machines are four-wheel-drive.

Backhoe

Backhoe. An attached implement on the rear of a tractor or crawler that is used to dig trenches. It is also referred to as a trencher.

Battery Electric, Battery Operated or Battery Powered. A miniature having a battery powered electric motor which provides power to operate it.

Blade. A front or rear mounted implement used to push or pull dirt, snow, manure, etc.

Bonnet. The covering of the engine. See also hood.

Bulldozer

Bulldozer. A crawler with a blade.

Cab. An enclosure over the operator's platform which protects the driver from the elements and from injury due to roll over. Many modern tractor cabs are equipped with a heater, air conditioner, radio, stereo and other options.

Cab

Canopy. A simple overhead protection on a machine designed to protect the operator from injury and/or the elements but not enclosing the operator as the cab does. It is usually part of a roll over protection system (ROPS).

Cast-in-driver

Cast-in-driver. The driver on some miniatures is cast as part of the machine.

Cast-in-name. The name on some cast metal models appears as raised lettering or as recessed lettering.

Cigarette Dispenser

Cigarette Dispenser. A miniature machine so made as to serve as a cigarette holder and dispenser.

Cigarette Lighter. A miniature machine with a part of it serving as a cigarette lighter.

Cleated Front Wheels. Front tires having traction lugs as found on front-wheel-assist and four-wheel-drive models.

Clockwork. A keywind mechanism which provides power for some miniatures. Some models have a ratchet mechanism which serves a similar purpose.

Collector Models or Collector Series

Collector Models or Collector Series. Models with special inscriptions made for collectors. Usually these models are made to commemorate the introduction of new models or special events.

Crawler—Loader

Crawler-Loader. A crawler with a front end hydraulic loader.

Culti-Vision

Culti-Vision. A term used by International Harvester Corporation to describe the offset design of some small tractors which aids in vision for cultivating row crops such as corn.

Custom or Custom made. Specially made models, usually in very low production numbers, not made on an assembly line.

Decal. the transparent paper or cellophane-like material bearing the trade name or model numbers used to identify miniatures. Some decals are water transfer while others are adhesive backed.

Decanter

Decanter. A miniature in the shape of a tractor but actually a bottle.

Disk. A farm implement having many wheel-like blades used to break up the soil after primary tillage.

Duals or Dual Rear Wheels

Duals or Dual Rear Wheels. Two wheels on the end of each axle providing greater traction and flotation on soft ground. Some four-wheel-drive tractors have duals on both the front and rear axles. See also triples.

Earth Mover. A machine used to move dirt from one location to another. Some are simply a trailer with a loading and/or unloading mechanism(s) while others are self propelled, having a power unit.

Electric. Some miniatures have an electric or battery electric power unit for propellsion.

Embossed. Having a stamped out design such as the grill on some models or the collector inscription on others.

Embossed

Exhaust Stack. The pipe, usually having a muffler, extending up through or along side of the hood transferring the exhaust away from the engine. Also called a silencer.

Fast Hitch

Fast Hitch. A term used by International Harvester to identify a tractor — implement hitching system consisting of two attachment points. It was popular during the 1950's.

First Edition. Early production models which are usually specially identified for collectors.

Flywheel

Flywheel. A heavy wheel at the end of a crankshaft which makes an engine run smoothly. On early one and two cylinder engines, the flywheel is usually exposed.

Four-Wheel-Drive (FWD)

Four-Wheel-Drive (FWD). A term used to describe a tractor that has power to both the front and rear wheels. All wheels are the same size. These tractors may also have dual or triple wheels on each axle.

Four-Wheel Skid Steer Loader. An implement used on farms and in industry for loading, scraping, etc. Both wheels on one side can be stopped while the other wheels keep turning causing the unit to "skid" around, hence, the name skid-steer.

Four-Wheel Skid Steer Loader

Four-Wheel-Steer or Crab Steer. All four wheels are used for steering permitting close manuvering or steering on steep hillsides. Generally this can be found only on four-wheel-drive machines.

Four-Wheel-Steer or Crab Steer

Front-Wheel-Assist. A term used to describe machines having power to the front as well as back wheels. These machines are similar to four-wheel-drive except the front wheels are smaller than the rear wheels. There are a few exceptions where the rear wheels are smaller than the front wheels.

Front-Wheel-Assist

Friction or Gyro Motor. A type of motor found on some miniatures.

Grill. A protective mesh like screen or series of bars designed to protect the radiator. Early tractors not having a grill were called "unstyled" while later tractors, particularly the late 1930's, having a grill were referred to as "styled".

Gyro Motor. A type of impulse motor used to power some miniatures. See also friction.

Half Tracks. A tractor having metal or rubber tracks attached to the rear wheels and an extra set of wheels, called "boogie" or idler wheels. These

Half-Tracks

tracks provided improved traction much like crawler tracks.

Hood. The covering of the engine. See also bonnet.

Hood

Hydraulic. A system comprised of a pump, lines, valves controls and a fluid used to transfer power from one point to another via hydraulic cylinders. The raising and lowering of implements is just one example of the use of hydraulics.

Hydro (Hydrostatic). A term used to describe a hydraulically operated transmission now found on certain tractors, etc. A transmission of this type replaces the conventional gear type and provides infinite foreward and reverse speeds.

Industrial

Industrial. Tractors and other machines used for non-farm uses such as construction equipment. These are usually yellow, some shade of yellow, orange, tan or some shade of brown.

Integral. A term used to describe the linking of a tractor and implement into one unit. An example is a tractor with a three-point-hitch plow.

Integral

Lawn and Garden Tractor (LGT). A small tractor used for mowing lawns and light garden work as well as many other light chores.

Loader. A bucket or scoop-like device on the front of a tractor used to load dirt, manure, etc.

Metal Wheels/Tires. Wheels and tires cast together as one piece and found on some miniatures particularly the cast metal models.

Mis-Labeled. An incorrect identification name or model number on a miniature.

Molded-In Driver. A driver molded into the tractor unit on either plastic or rubber miniatures.

"Mod" Tractor

"Mod" Tractor. A miniature that is not realistic in appearance but rather a modernistic creation.

Original. The first of that particular model in miniature.

Paperweight. A solid cast model, usually not having a great amount of detail, designed primarily as a desk piece.

Plastic Wheels. The wheels on a miniature being made of plastic rather than metal. Frequently the wheels are molded with the tires in one piece.

Plated. A part, usually metal, having a surface coating of a shiny material, such as chrome or nickle. Some plastic parts have a similar "plating".

Radiator. The cooling mechanism of an engine. Some engines have no radiator and, therefore, are said to be air cooled.

Recast. A model made years later from the original molds or dies. These are not usually as valuable as the originals.

Remote Control

Remote Control. The device attached to some miniatures which permits steering and directional control. These are usually electric powered models.

Reproduction. A copy of the original model. These are worth much less than originals.

Roll-Over-Protection-System (ROPS). A frame like attachment to a tractor designed to protect the operator from injury in the event of roll over. The ROPS can be combined with a canopy or incorporated into a cab to provide the operator with additional protection from the elements.

Roto-Tiller. An implement which breaks up soil using a series of tines.

mounted on the rear of the tractor or implement providing oncoming drivers of a vehicle travelling less than twenty-five miles per hour. Many later miniatures have miniature SMV signs mounted on the rear.

Sound Gard Cab. A trademark name given to a safety cab designed by engineers of Deere and Company.

Row Crop. A farm tractor having a single or two closely spaced front wheels which provide for easy cultivation of row crops.

Rubber Tracks. Used on many miniature crawler models.

Rubber Wheels/Tires. Miniatures having the wheels and tires molded as a single piece.

Safety Frame. A type of roll-over-protection-system (ROPS).

Standard. A term used to describe a style of tractor having a fixed wheel tread or axle spacing and a low center of gravity. This is different than the row crop style tractor.

Separate Driver. Older miniatures having drivers frequently had the drivers cast separately. In many cases, these drivers are nickle plated.

Slow Moving Vehicle Sign (SMV). A safety emblem with both daylight and night reflective material

Steam Traction Engine. A very old style tractor powered by a steam boiler type of engine. This style tractor was first an adaptation of a railroad locomotive fitted with lug type wheels.

"Steel". A term used to describe a type of wheel used on early farm tractors before rubber tires became popular, and still used today by certain religous groups who farm. With miniatures, this style wheel is not necessarily made of steel but could be diecast metal, plastic or wood.

"Steel"

Super Rod. A term used to describe some versions of pulling tractors.

Three-Point-Hitch. An integral hitching system used on most modern farm tractors consisting of two lower lifting links and an upper stabilizing link. Most tractors have the three-point-hitch on the rear but some also have this system on the front.

Three-Point-Hitch

Trailer. A two or four wheel implement used for hauling various items. There are many types of trailers including rear dump, side dump, gravity, forage, bale wagons, etc.

Trencher. A type of backhoe.

Triples

Triples. Three wheels on the end of each axle.

Turf Tires. Specially treaded tires which provide traction but do not tear up the turf as much as conventional lug type tires. This type of tires are frequently found on lawn and garden tractors.

Turf Tires

Turbo (Turbocharger). A blower on a diesel engine which provides more air for greater engine efficiency and greater power.

Utility

Utility. A type of tractor having a wide front axle and low center of gravity used for many farm and industrial chores.

V-Blade. A front mounted blade shaped like a V that pushes dirt or snow off to both sides.

Weights

Weights. Extra ballast used for greater traction. Weights can be added on the front, sides, or rear of the tractor.

Wheatland. A term used to describe a style of tractor which is used on the great plains. See also standard.

Wide Front End (WFE)

Wide Front End (WFE). Tractor front wheels spaced the same or nearly the same distance apart as the rear wheels. This style provides greater tractor stability, particularly on hillsides.

Winch. A device on the front or rear of a tractor or crawler used to pull loads such as logs. It can also be used to pull out a "stuck" tractor by anchoring it to a solid object and pulling itself out.

Wood Wheels. Some older cast iron miniatures had wooden wheels/tires rather than the separate wheels and rubber tires.

2 + 2. International Harvester's designation for their four-wheel-drive row crop tractors.

Pedal Tractor Glossary

Rear Axle

Axle, rear with sprocket. Left—Current type, using regular 1'' pitch 3/8'' bicycle chain. Axle diameter 1/2''. Right—Early Eska-7/16'' shaft with "skip" tooth sprocket with matching chain of #61-3/16 twin roller type. Note: One end of axle shaft has flattened surface to allow for engagement of "driver" wheel.

Bearings and bushings. Left—Current plastic bushing used for rear axle. Right—Metal flange, ball bearing used on earlier pedal crank supports.

Hitches

Hitches. Early "Eska" hitch like found on small "H" Farmall tractor. A broad flat hitch area with small hole for attaching trailing implements.

Hitches

Hitches. Latch type locking devise used on many "Eska" made models. Lifting up on latch releases trailer hitch. Note hitch pattern shown below trailer connection.

Hitches

Hitches. A later "Eska" type hitch utilizing a "draw-pin" to attach trailer.

Hitches

Hitches. This hitch features a large loop which fastened to cast in projection at rear of pedal tractor. Trailer tongue is also bent to accomodate the higher hitch position.

Hubs

Hubs. Left—"Dome" or "sperical" shaped hub cap used on early "Eska" models. Painted same color as wheels. Center—Current, plated type hub cap. Flatter design than Dome type. Right—Current push-nut to hold "Ertl" plastic front wheels on spindles.

Noise Maker

Noise maker. Flexible spring steel strip fastened to a lever. Pushing this lever caused the flexible steel strip to contact sprocket teeth, causing a simulated engine noise.

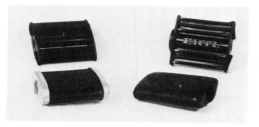

Pedals

Pedals. Top row: Left—Plastic without "Ertl" name molded on it. Used on early "Ertl" selected models. Also made in white for models-AC D-17. Right—Current "Ertl" style. Plastic with "ERTL" name molded into pedal. Bottom row: Left—Metal end "teardrop" style used on "Eska" models. Right—Solid rubber without metal ends. Used on very early Eska models.

Pedals

Pedals. "Tear drop", metal end type pedal on a John Deere "620" type pedal tractor.

Pedals

Pedals. The solid rubber pedal used on very early "Eska" made models. Metal ends were not used on this style.

Seats

Seats. Top row: Left—Cast aluminum, used on early Eska models. Right—Stamped steel used on most later "Eska" models. Note three parallel indentation on seat base. Bottom row: Left—Metal diecast type used on Ertl tractors. Right—Cast aluminum used on Ford 900. Middle—Stamped steel, used on Ford 6000 Commander.

Seats. Top row: Left—Cast aluminum—used only on early "Eska" John Deere "A". Right—Cast aluminum—used on John Deere 60 thru 630 type.

Seats

Bottom row: Left—Die-cast, used on John Deere 4020, 4010 models. Note the raised back rest design. Right—Current molded plastic type.

Trailers

Trailers. An early type trailer with fenders. All metal construction. Used with latch-type hitching devise on tractor. Metal wheels with hub caps. Available in all major tractor manufacturers colors with trade mark or brand name on rear of trailer.

Trailers

Trailers. Later type trailer using plastic wheels and no fenders.

Trailers

Trailers. Hitch variation, with large loop on end to accomadate the hitching point on tractor. This trailer has metal wheels with cleated tires and hub caps.

Trailers. This new styled stake trailer from Ertl features all plastic construction with side boards to keep load in place. It is designed to be hitched to the new plastic International tractors. Measures 21 inches long.

400AO NEW

Rear Wheel & Tire

Wheel and tire, rear. Left–Rubber tire on metal rim with raised spokes design. Used on many "Eska" models. Wheel contained 3 slots to accept hub cap. "Eska-Dubuque, Iowa" molded in tire. 12 x 1.75. Right—Current type metal wheel with rubber tire. "Ertl" molded in tire. Note: *Driver wheel*— has rectangular hole on outer side to accept flat end of axle to give drive connection. *Idle wheel*— Round hole, does not produce a driving action.

Wheels, steering. Left—Small $5\frac{7}{8}$" diameter, used only on early "Eska" John Deere "A". Cast aluminum. Used set screw to fasten to steering shaft. Right—Round tubular steel with flat steel bar. 7" diameter. Used on early "ESKA" models.

Steering Wheels

Wheels, steering. Left—$6\frac{7}{8}$" cast aluminum, used on many "Eska" models. Used set screw to fasten to steering shaft. Right—8" die-cast, with finger grips. Hollow underside of outer rim.

Front Wheels & Tires

Steering Wheels

Wheels and tires, Front. Top row: Left—Current plastic wheel with $1\frac{1}{2}$" wide tire—Ertl. Middle— Metal spoke design with rubber tire marked Eska Dubuque Iowa $1\frac{1}{4}$" wide tread. Right—Narrow 1" x 6" tire like used Case VAC. Bottom row: Left—Metal wheel with $1\frac{1}{2}$" rubber tire. No spoke design. Designed for use with hub cab. Middle— Smooth treadless tire marked "Eska Company Inc. Dubuque, Iowa 6 x 1:00". Right—"ORCO Ohio Rubber Company, Willoughby, Ohio 7.00 x 1.25 Air King" molded into tire. Used on some early pedal tractors. Not pictured: a) Some pedal tractors may be found with "Hamilton" name on front tire, also made by Ohio Rubber Co., Willoughby Ohio, size 7.00 x 1.25. b) Front tire used on small Massey Harris 44 "Swan, 7 x 1.25, puncture proof" molded in rubber.

Wheel, steering. Current "Ertl" plastic style having $7\frac{7}{8}$" diameter. Thicker construction than metal steering wheels.

Index